A THEORY
OF MODULATION

EINE
MODULATIONSTHEORIE

Da Capo Press Music Reprint Series

GENERAL EDITOR

ROLAND JACKSON

A THEORY
OF MODULATION

EINE
MODULATIONSTHEORIE

THORVALD OTTERSTRÖM

DA CAPO PRESS • NEW YORK • 1975

Library of Congress Cataloging in Publication Data

Otterstrom, Thorvald, 1868-
  A theory of modulation Eine Modulations- theorie.

  (Da Capo Press music reprint series)
  Reprint of the ed. published by University of Chicago
Press, Chicago.
  1. Modulation (Music) 2. Musical intervals and
scales. 3. Permutations. 4. Music and color.
I. Title. II. Title: Eine Modulationstheorie.
MT52.088   1975   781.3   74-34379
ISBN 0-306-70721-7

This Da Capo Press edition of *A Theory of Modulation/Eine Modulationstheorie* is an
unabridged republication of the first edition published in Chicago in 1935. It is reprinted
from an original in the collections of the Memorial Library, University of Wisconsin.

Copyright 1935 by The University of Chicago.

Published by Da Capo Press, Inc.
A Subsidiary of Plenum Publishing Corporation
227 West 17th Street, New York, N.Y. 10011

A THEORY OF MODULATION

EINE MODULATIONSTHEORIE

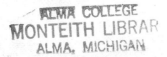

THE UNIVERSITY OF CHICAGO PRESS
CHICAGO, ILLINOIS

———

THE BAKER & TAYLOR COMPANY
NEW YORK

THE CAMBRIDGE UNIVERSITY PRESS
LONDON

THE MARUZEN-KABUSHIKI-KAISHA
TOKYO, OSAKA, KYOTO, FUKUOKA, SENDAI

THE COMMERCIAL PRESS, LIMITED
SHANGHAI

THORVALD OTTERSTRÖM

A THEORY
OF MODULATION

EINE
MODULATIONSTHEORIE

THE UNIVERSITY OF CHICAGO PRESS
CHICAGO · ILLINOIS

THE UNIVERSITY OF CHICAGO PRESS
CHICAGO, ILLINOIS

———

THE BAKER & TAYLOR COMPANY
NEW YORK

THE CAMBRIDGE UNIVERSITY PRESS
LONDON

THE MARUZEN-KABUSHIKI-KAISHA
TOKYO, OSAKA, KYOTO, FUKUOKA, SENDAI

THE COMMERCIAL PRESS, LIMITED
SHANGHAI

ERRATA

Page 67, bar 47, bass: *for* b *read* b flat
Page 83, bar 7, soprano: *for* g *read* g sharp
Page 93, bar 9, bass: *for* c *read* d
Page 95, bar 14, last chord: × in
Page 95, bar 15, last chord: × out
Page 95, bar 20, soprano: *for* a *read* c
Page 98, bar 18, bass: *for* a double flat *read* b double flat
Page 103, bar 2, soprano: *for* d sharp *read* d
Page 108, bar 7, piano, right hand, inner part: *for* g g *read* g sharp g
Page 119, bar 12, beat 1, soprano: *for* d *read* e

DRUCKFEHLER

Seite 67, Takt 47, Bass: *anstatt* h *lese* b
Seite 83, Takt 7, Sopran: *anstatt* g *lese* gis.
Seite 93, Takt 9, Bass: *anstatt* c *lese* d
Seite 95, Takt 14, Zweiter Akkord: × ein
Seite 95, Takt 15, Zweiter Akkord: × weg
Seite 95, Takt 20, Sopran: *anstatt* a *lese* c
Seite 98, Takt 18, Bass: *anstatt* asas *lese* bb
Seite 103, Takt 2, Sopran: *anstatt* dis *lese* d
Seite 108, Takt 7, Piano, rechte Hand, Mittelstimme: *anstatt* g g *lese* gis g
Seite 119, Takt 12, Sopran, erstes Viertel: *anstatt* d *lese* e
Seite 155, unter *"Des dur"*: blau-orange *lese* blau-violett

PREFACE

In this work it has been attempted to develop chords and modulations on a mathematical basis.

It has been known for centuries that the major third lies below in a major triad and above in a minor triad. It has perhaps been less known or observed that a mathematical process—permutation—has taken place and that the two chords belong in the same class because their factors are alike. If the thirds are expressed in numbers (these representing a measurement in half-tones), this fact becomes clearer. The number 4 represents a major third, 3 a minor third; 43 (to be read: "four three") is a major triad, 34 a minor triad; all possibilities have been exhausted with this combination according to the laws of permutation.

In the seventh-chords (three thirds) there are three classes: 334, 344, 333. The first two classes produce three chords each; the third class admits of no permutation. By this means every chord has been identified with a number form, which at the same time explains its construction. This is an advantage over the traditional way of classifying these chords in primary and secondary seventh-chords, since the latter group does not give proper identification to each chord, and, besides, is misleading if the term "secondary" is understood as secondary in importance. All chords are of equal importance.

The confusion is greater in regard to chromatically altered seventh-chords, of which only two are generally known (in the textbooks). But there are nine; the chord elements are 424 with three possibilities and 234 with six.

As far as known to this author, no attempt has been made to develop ninth-chords systematically; only a few—perhaps six—have been recorded. It has been demonstrated in this work that there are seventy. The index shows the groups and the development possibilities of each group. Some theorists assert that the fourth inversion of the ninth-chord is absurd and cannot be used. The examples in this book show that every ninth-chord can be used in all four inversions.

The modulation forms produce key connections upon a mathematical basis—partition. One result of this is the elimination of sequences.

Of seven-tone scales we have known the seven ecclesiastical scales, the major scale, and the three minor scales (the harmonic minor, the melodic minor, and the Hungarian minor)—together eleven scales; as the Ionian scale and the major scale are identical, the number is ten. It is here shown that the permutation of these scales produces two hundred and sixty-six scales.

The relation of tone and spectrum color discussed in these pages is merely a suggestion; no claim for its scientific accuracy is made.

The author gratefully acknowledges the assistance of his pupils, B. Royt and N. Lupu, and also of Dr. Ernest Bloomfield Zeisler, whose mathematical treatment will be found in the Appendix.

CHICAGO
1935

VORWORT

In diesem Werke ist der Versuch gemacht, Akkorde und Modulationen auf mathematischer Grundlage zu entwickeln.

Seit Jahrhunderten ist bekannt, dass in einem Durdreiklange die grosse Terz unten und in einem Molldreiklange oben liegt. Weniger wahrscheinlich wurde beobachtet, dass ein mathematischer Vorgang—Permutation—stattfand und dass die beiden Akkorde zu einer Klasse gehören, da ihre Bestandteile gleich sind. Wenn die Terzen in Zahlen dargestellt werden (welche in Halbton-Messung ausgedrückt werden), so wird die Sache klarer. 4 bedeutet eine grosse Terz, 3 eine kleine; 43 (lies: *vier drei*) ist ein Durdreiklang, 34 ein Molldreiklang; nach der Formel der Permutation sind mit dieser Kombination alle Möglichkeiten erschöpft.

Die Septimenakkorde (drei Terzen) bestehen aus drei Klassen: 334, 344, 333. Die beiden ersten Klassen ergeben je drei Akkorde; die dritte Klasse ist nicht weiter entwicklungsfähig. Auf diese Weise wird jeder Akkord durch eine Zahlenformel, welche die Konstruktion erklärt, bezeichnet. Gegenüber der herkömmlichen Art der Klassifizierung in Haupt- und Nebenseptimenakkorde ist dies ein Fortschritt, da die alte Weise nicht jeden Akkord genau bezeichnet, auch zweideutig ist, indem diese Akkorde als untergeordnet hingestellt werden. Alle Akkorde sind von gleicher Wichtigkeit.

Die Verwirrung ist noch grösser in Bezug auf die alterierten Septimenakkorde, von welchen bisher nur zwei allgemein bekannt waren (in Lehrbüchern). Doch giebt es deren neun: 424 mit drei und 234 mit sechs Möglichkeiten.

Soweit dem Verfasser bekannt, ist bisher keine wissenschaftliche Entwicklung der Nonenakkorde unternommen worden; nur wenige—vielleicht sechs—wurden erwähnt. In diesem Werke ist bewiesen, dass deren Zahl 70 beträgt. Das Verzeichnis führt jede Gruppe und ihre Entwicklungsmöglichkeiten an. Gewisse Theoretiker bestreiten die Möglichkeit der vierten Umkehrung der Nonenakkorde. Die Beispiele dieses Werkes beweisen das Gegenteil.

Die Modulationsformel ermöglicht Verbindungen auf dem mathematischen Vorgang der Verteilung ("Partition"); infolgedessen wird die Sequenz ausgeschaltet.

Von den siebenstufigen Tonleitern kennen wir (1) die sieben Kirchentöne, (2) die Durskala, (3) die drei Mollskalen (harmonische, melodische, und ungarische)—zusammen elf Skalen. Da die jonische und Durskala gleich sind, ist deren Zahl nur zehn. Der Beweis ist geliefert, dass 266 Skalen durch Permutation möglich sind.

Das Verhältnis von Ton und Spectral-Farbe (in diesem Werke erwähnt) soll nur als Andeutung von Möglichkeiten betrachtet werden, indem es keinen Anspruch auf wissenschaftliche Genauigkeit macht.

Der Verfasser ist seinen Schülern B. Royt und N. Lupu für freundliche Mitwirkung zu herzlichem Danke verpflichtet, sowie Dr. Ernest Bloomfield Zeisler für die mathematische Darstellung im Nachtrag.

CHICAGO
1935

CONTENTS

	PAGE
INDEX	viii
MODULATION	1
MODULATION MODELS	7
SELECTION OF KEY	20
KEY-FORMS	23
SYMMETRY I	41
SYMMETRY II	43
APPLICATION OF KEY-FORMS	44
SEVENTY NINTH-CHORDS DEVELOPED BY MEANS OF PERMUTATION	59
A MODULATION EXAMPLE FURTHER DEVELOPED	102
MODULATION FORMS AND THE SYMMETRIC INVERSION	113
THE DIMINISHED SEVENTH-CHORD	123
ON SCALES	127
PERMUTATION APPLIED TO SCALES	130
PERMUTATION OF THE MAJOR SCALE	134
PERMUTATION OF THE HARMONIC MINOR SCALE	135
PERMUTATION OF THE HUNGARIAN SCALE	139
NEW CHORDS	144
A NEW SCALE DEVELOPED AS PIANO STUDY	146
A SUGGESTION OF HARMONIES TO THIS SCALE	151
THE SPECTRUM COLORS AND THE MODULATION FORMS	152
APPENDIX BY DR. ERNEST B. ZEISLER	156

INHALT

	SEITE
INDEX	viii
MODULATION	1
MODULATIONSMUSTERN	7
WAHL DER TONART	20
GRUND-FORMELN	23
SYMMETRIE I	41
SYMMETRIE II	43
NUTZANWENDUNG DER GRUND-FORMELN	44
SIEBZIG NONENAKKORDE ENTWICKELT DURCH PERMUTATION	59
EIN MODULATIONSBEISPIEL WEITER ENTWICKELT	102
MODULATIONSFORMELN UND DIE SYMMETRISCHE UMKEHRUNG	113
DER VERMINDERTE SEPTIMENAKKORD	123
ÜBER SKALEN	127
PERMUTATION AUF SKALEN ANGEWANDT	130
PERMUTATION VON DER DUR-SKALA	134
PERMUTATION VON DER HARMONISCHEN MOLL-SKALA	135
PERMUTATION VON DER UNGARISCHEN SKALA	139
NEUE AKKORDE	144
EINE NEUE SKALA ALS KLAVIERÜBUNG ENTWICKELT	146
EINE ANDEUTUNG ZUR HARMONISIERUNG DIESER SKALA	151
DIE SPECTRALFARBEN UND DIE MODULATIONSFORMELN	152
ANHANG VON DR. ERNEST B. ZEISLER	156

INDEX OF NINTH-CHORDS
VERZEICHNISS DER NONENAKKORDE

	PAGE SEITE		PAGE SEITE
2444	78	3444	66
4244	61	4344	67
4424	65	4434	67
4442	77	4443	66
3334	90	2345	83
3343	89	2354	71
3433	88	2435	79
4333	60	2453	75
		2534	74
2335	97	2543	72
2353	98		
2533	97	3245	62
3235	96	3254	73
3253	96	3425	70
3325	95	3452	72
3352	95	3524	81
3523	94	3542	75
3532	94		
5233	93	4235	63
5323	93	4253	81
5332	92	4325	68
		4352	74
3335	91	4523	73
3353	80	4532	71
3533	80		
5333	82	5234	76
		5243	76
3344	87	5324	69
3434	86	5342	98
3443	84	5423	77
4334	60	5432	69
4343	85		
4433	64		
2344	83		
2434	79		
2443	78		
3244	62		
3424	70		
3442	84		
4234	63		
4243	61		
4324	68		
4342	82		
4423	65		
4432	64		

MODULATION

Modulation is a transition to another key. Two factors are required to produce a modulation: selection of key and the medium whereby this is reached. The selection of key may be arbitrary and may be produced by arithmetic forms. These will be discussed later.

Any chord may serve as means of reaching another key. A major triad may belong to six different keys and can be used as a medium of modulation to these six tonalities. A major triad on C is tonic in C major, dominant in F major, subdominant in G major, dominant in F minor, submediant in E minor, and Neapolitan sixth in B minor.

The term "Neapolitan sixth" was probably used for the first time in 1868 by L. Bussler, who attributed this chord to "the famous composer Scarlatti." To us the "famous composer Scarlatti" is Domenico Scarlatti; but he does not use this chord, for he was a "modernist" and spurned a chord which was very common. The "famous composer," therefore, must have been his father, Allessandro Scarlatti, who was born in Sicily in 1659. He lived for a while in Naples, and this is the only reason for the geographical appellation "Neapolitan"! The probability is that this chord was taken over from the Phrygian scale, since the second degree in this minor scale is one half-tone above the tonic and has a major triad.

MODULATION

Unter Modulation versteht man den Übergang aus einer Tonart in eine andere. Zwei Faktoren sind notwendig um eine Modulation zu erlangen: erstens die Wahl der Tonart; zweitens die Mittel, durch die man diesen Übergang vollbringt. Die Wahl der Tonart kann willkürlich sein oder sie kann durch arithmetische Formeln bedingt werden. Diese werden später erörtert.

Jeder beliebige Akkord kann als Übergang in eine andere Tonart verwendet werden. Ein Durdreiklang kann sechs verschiedenen Tonarten angehören und kann als Mittel zur Modulation nach diesen sechs Tonarten gebraucht werden. Ein Durdreiklang auf C ist erste Stufe in C Dur, fünfte Stufe in F Dur, vierte Stufe in G Dur, fünfte Stufe in F Moll, sechste Stufe in E Moll und Neapolitanische Sexte in H Moll.

Die Bezeichnung Neapolitanische Sexte ist wahrscheinlich zum ersten Male im Jahre 1868 von L. Bussler angewandt worden, der den Ursprung dieses Akkordes dem "berühmten Komponisten" Scarlatti zuschreibt. Dieser "berühmte Komponist" ist für uns Domenico Scarlatti. Er hat aber diesen Akkord nicht angewandt; er war nämlich ein Modernist und verschmähte einen so gewöhnlichen Akkord. Der "berühmte Komponist" muss sein Vater, Allessandro Scarlatti sein, der in Sicilien in 1659 geboren war. Er lebte eine Zeitlang in Neapel, und das ist der ganze Grund für die geographische Benennung! Wahrscheinlich ist dieser Akkord von der Phrygischen Tonleiter übernommen, denn die zweite Stufe dieser Tonleiter ist eine halbe Stufe höher als die Tonika und hat einen Durdreiklang.

The major triad Der Durdreiklang

The minor triad can belong to five different keys; a minor triad on D is supertonic in C major, mediant in B flat major, submediant in F major, tonic in D minor, and subdominant in A minor.

Der Molldreiklang kann fünf verschiedene Tonarten angehören; ein Molldreiklang auf D ist zweite Stufe in C Dur, dritte Stufe in B Dur, sechste Stufe in F Dur, Tonika in D Moll und vierte Stufe in A Moll.

The most interesting feature of the major triad as a medium of modulation is that it furnishes direct connection between two minor keys a tritone apart. It is dominant in one and Neapolitan sixth in the other. Thus a C major triad is dominant in F minor and Neapolitan sixth in B minor. By the application of

Das interessanteste Merkmal des Durdreiklanges als Modulationsmittel ist, dass er eine direkte Verbindung zwischen zwei Molltonarten darstellt, die in Tritonusverhältnis zu einander stehen. Dieser Akkord ist in der einen Tonart Dominante, in der anderen Neapolitanische Sexte. So ist also ein C Dur Dreiklang Dominante in F Moll und Neapolitanische Sexte in H Moll. Wendet man das Princip der symmetrischen

the principle of symmetric inversion a minor triad connects two major keys a tritone apart.

Umkehrung an, so verbindet ein Moll-dreiklang zwei Dur Tonarten, die in Tritonusverhältnis zu einander stehen.

The augmented triad has twelve resolutions produced by half-tone moves: each tone may be moved one half-tone up or down; the lower third and the upper third may be moved one half-tone up or down; and the two outside tones may move one half-tone up or down. Enharmonic changes are, of course, necessary.

Der übermässige Dreiklang hat zwölf Auflösungen, die durch Halbtonfortschreitungen hervorgebracht werden: jeder einzelne Ton kann einen halben Ton aufwärts oder abwärts sich bewegen. Die untere Terz und die obere Terz können einen halben Ton aufwärts oder abwärts schreiten; ebenso die beiden Aussentöne einen halben Ton aufwärts oder abwärts. Enharmonische Veränderungen sind natürlich erforderlich.

The diminished seventh-chord may be changed enharmonically and located in four different keys.

Der verminderte Septimenakkord enharmonisch verändert kann vier verschiedenen Tonarten angehören.

In the following models for modulation seventh-chords are used; and later these chords are used again, each with a major ninth and a minor ninth. For the sake of convenience, each chord is expressed in a form of numbers, each number indicating an interval measured in half-tones —e.g., 333. To a given tone add a minor third (3); to this add another minor third (33); then still another minor third (333). All moves are, of course, upward. The number-form 442 means a major third, again a major third, and a diminished third. In every chord, by adding these numbers, the size of each interval, figured from the root, can easily be recognized: i.e., in the first chord (333) the third is minor (3), the fifth is diminished (3+3), and the seventh is diminished (3+3+3). In the second chord (442) the third is major (4), the fifth is augmented (4+4), and the seventh is minor (4+4+2). The term "secondary seventh-chords" applied collectively to five of the seven diatonic seventh-chords is not practical because of the confusing lack of distinct identification. Every chord might have a name of two adjectives, the first describing the triad and the second the seventh: e.g., "diminished minor" (334), "small minor" (343), "large major" (434), "augmented major" (443), and "small major" (344). But a form of numbers is simpler and more direct than cumbersome names.

With the nine chromatic seventh-chords, the confusion is still greater. Ziehn gave each one of these chords a Roman numeral; but another theorist might wish to arrange the chords in a different, more practical order, and the resulting two or more arrangements would create only confusion for the student. But there can be no confusion or difference of opinion when the structure of the chords is designated by numbers.

In den folgenden Modulationsmustern sind Septimenakkorde verwendet, und später einzeln wieder gebraucht mit grosser und kleiner None. Der Bequemlichkeit halber wird jeder Akkord mit einer Zahlenformel bezeichnet, so dass jede Zahl ein in Halbtönen gemessenes Intervall andeutet, z.B. 333: einem gegebenen Ton fügt man eine kleine Terz zu (3), dann wieder eine kleine Terz (33) und dann nochmals eine kleine Terz (333). Alle Bewegungen sind aufwärts gemeint. Die Zahlenformel 442 bedeutet eine grosse Terz, nochmals eine grosse Terz und eine verminderte Terz. Wenn man diese Nummern addiert, kann man in jedem Akkord die Grösse der einzelnen Intervalle, vom Grundtone aus gerechnet, leicht erkennen: das heisst, in dem ersten Akkord ist die Terz klein (3), die Quinte vermindert (3+3) und die Septime ebenfalls vermindert (3+3+3). In dem zweiten Akkorde ist die Terz gross (4), die Quinte übermässig (4+4) und die Septime klein (4+4+2). Die Bezeichnung Neben-Septimenakkorde, als Kollektivname für fünf von den sieben diatonischen Septimenakkorden angewandt ist unpraktisch wegen der Unbestimmtheit. Mit der Bezeichnung eines jeden Akkordes könnte man zwei Adjektive verbinden, von denen das erste den Dreiklang und das zweite die Septime beschreibt, z.B. vermindertes Moll (334), kleines Moll (343), grosses Dur (434), übermässiges Dur (443), kleines Dur (344). Eine Zahlenformel ist aber einfacher und direkter als schwerfällige Bezeichnungen.

Bei den neun alterierten Septimenakkorden ist die Verwirrung noch grösser. Ziehn hat jeden von diesen Akkorden mit einer römischen Zahl bezeichnet. Man könnte vielleicht diese Akkorde in eine andere, mehr praktische Anordnung bringen, und das Resultat zweier oder vielleicht noch weiterer Anordnungen würde für den Schüler nur Verwirrung sein. Es kann aber keine Verwirrung oder Meinungsverschiedenheit entstehen wenn der Bau der Akkorde durch eine Zahlenformel ausgedrückt ist. Wenn wir

When we see the form 324, we can at once construct the chord: minor third, diminished third, major third; and it is of no importance whether this chord is labeled "II," "III," "V," or "VIII."

die Formel 324 sehen, können wir sofort den Akkord herstellen: kleine Terz, verminderte Terz, grosse Terz, und es ist nebensächlich, ob dieser Akkord als II, III, V oder VIII bezeichnet wird.

GENERAL VIEW OF INTERVALS TO NINTH CALCULATED IN HALF-TONES

ALLGEMEINE ÜBERSICHT DER INTERVALLE BIS ZUR NONE, IN HALBTÖNEN GERECHNET

The starting-point is 0 and is, of course, optional.

Der Anfangspunkt ist willkürlich (kann als 0 bezeichnet werden).

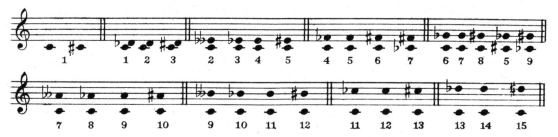

SEVENTH-CHORDS IN NUMBER-FORMS AND THEIR PLACE IN THE SCALE

SEPTIMENAKKORDE IN ZAHLENFORMELN UND DEREN STELLUNG IN DER TONLEITER

333	VII in minor (and major)	VII in Moll (und Dur)
334	VII in major and II in minor	VII in Dur; II in Moll
343	II, III, VI in major; IV in minor	II, III, VI in Dur; IV in Moll
433	V in major and V in minor	V in Dur und V in Moll
434	I, IV in major; VI in minor	I, IV in Dur; VI in Moll
443	III in minor	III in Moll
344	I in minor	I in Moll
424	II in minor with raised third	II in Moll, Terz erhöht
324	VII in minor with flatted fifth	VII in Moll, Quinte herabgesetzt
423	VII in minor with raised third	VII in Moll, Terz erhöht
442	V in major with raised fifth	V in Dur, Quinte erhöht
342	VII in minor with raised fifth	VII in Moll, Quinte erhöht
432	V in minor with flatted seventh	V in Moll, Septime herabgesetzt
244	II in minor with flatted third	II in Moll, Terz herabgesetzt
234	II in minor with raised root	II in Moll, Grundton erhöht
243	IV in minor with raised root	IV in Moll, Grundton erhöht

In the following models the modulation progresses one half-tone up and then returns to the starting-point. It is the assignment of the student to make modulations to all other keys. If the starting-point is C, the modulation chord is to be placed in a suitable position and then approached, care being taken to remain in the old key as long as possible and to avoid cadences to other keys. Chromatic

In den folgenden Mustern bewegt sich die Modulation einen halben Ton aufwärts und dann zurück nach dem Anfangspunkte. Es ist die Aufgabe des Schülers, durch Modulationen in alle anderen Tonarten zu gelangen. Wenn der Anfangston C ist, sollte der Modulationsakkord in angemessener Lage erscheinen; auf diese richtet sich das Ziel, unter möglichster Beibehaltung der An-

passing tones and suspensions may be used in such a manner that they do not establish another key than C major (or minor). When the new key has been reached and established satisfactorily, the return should be made to the first key in a similar manner.

If the examples are worked out in skeletal chords, it is recommended that the period of eight bars be observed; but it may be suggested to work out the models in figuration and in different time ($^3/_4$, $^6/_8$, $^9/_8$, $^5/_4$, etc.), as exercises in form and meter.

fangstonart und unter Vermeidung von Kadenzen nach anderen Tonarten. Chromatische Durchgangstöne und Vorhalte können in solcher Weise gebraucht werden, dass sie keine andere Tonart als C Dur (oder Moll) berühren. Wenn die neue Tonart erreicht und in zufriedenstellender Weise festgestellt ist, geht der Rückgang in die erste Tonart in ähnlicher Weise vor sich.

Für ein Muster in Akkorden ohne jegliche Figuration empfiehlt sich die Periode von acht Takten. Auch Muster in Figuration und verschiedenen Taktarten sind durchzuarbeiten ($^3/_4$, $^6/_8$, $^9/_8$, $^5/_4$, etc.), als metrische- und Formübungen.

SELECTION OF KEY

In the preceding models only one modulation was made with each chord one half-tone up, the task of the student being to modulate to the remaining ten keys with the same chord. The finished work will result in eleven separate exercises. It will now be investigated whether this amount of work can be concentrated into one exercise, whereby the student may save time and yet have the experience of modulating to all keys. The problem is this: ALL KEYS MUST BE REACHED WITHOUT REPEATING ANY AND WITHOUT REPEATING THE DISTANCE BETWEEN ANY TWO ADJACENT KEYS.

If the modulation has been made from C to E, we cannot move from F to A, from A flat to C; or, generally speaking, we cannot twice modulate a major third up. To solve this problem we will use the numbers from 0 to 11. Our whole modulation problem lies on the face of a watch, except that we substitute 0 for 12, which is our starting-point; from this we must reach all the other numbers without repeating any, and without repeating any move from one number to another.

We will first arrange the numbers in this order:

1 4 3 2 5 6 7 10 9 8 11

Then by addition we produce another form; in the lower line, somewhat to the left, we write 0, thus:

  1 4 3 2 5 6 7 10 9 8 11
0

Lower and upper lines are now added diagonally. (The base of addition is 12,

WAHL DER TONART

In den bisherigen Mustern ist nur eine Modulation mit jedem Akkorde einen halben Ton nach oben ausgeführt worden und es wurde dem Schüler die Aufgabe gestellt, nach den übrigen zehn Tonarten mit demselben Akkord zu gelangen. End-Resultat: elf verschiedene Übungen. Es soll nun untersucht werden ob diese Gedankenarbeit in *eine* Übung zusammengefasst werden kann, so dass der Schüler Zeit sparen und trotzdem die Erfahrung gewinnen konnte, in alle Tonarten zu modulieren. Das Problem ist folgendes: *Alle Tonarten müssen erreicht werden, ohne irgend eine zu wiederholen, und auch ohne den Abstand zwischen irgend-welchen zwei benachbarten Tonarten zu wiederholen.*

Wenn die Modulation von C nach E gemacht worden ist, können wir uns nicht von F nach A bewegen, auch nicht von As nach C, oder allgemein ausgedrückt: wir können nicht wiederum eine grosse Terz nach oben modulieren. Um diese Aufgabe zu lösen werden wir uns der Nummern von 0 bis 11 bedienen. Unser ganzes Modulationsproblem lässt sich auf dem Zifferblatte einer Uhr verfolgen; 0 ist gleich 12 und ist unser Anfangspunkt; von hier aus müssen wir alle anderen Nummern erreichen ohne eine zu wiederholen und ohne eine Bewegung von einer Nummer zu einer anderen zu wiederholen.

Wir werden erst die Nummern in dieser Ordnung aufstellen:

1 4 3 2 5 6 7 10 9 8 11

Dann produzieren wir durch Addition eine andere Formel; in der unteren Reihe schreiben wir 0:

  1 4 3 2 5 6 7 10 9 8 11
0

Obere und untere Reihe werden nun diagonal addiert. (Die Basis der Addi-

as in adding hours: 12=0; 13=1; 14=2; etc.)

```
  →1 →4 →3 →2  5  6  7  10  9  8  11
0   1  5  8  10  3  9  4   2 11  7   6
```

In the lower line we have reached all eleven numbers from the starting-point 0. The numbers in the upper line control the moves in the lower line: from 0 to 1 is 1; from 1 to 5 is 4; from 5 to 8 is 3; etc. No number is repeated in either line, and therefore we know that our problem is solved. The upper line we will call the "key-form"; the lower line the "work-form."

Each key-form produces four work-forms:

```
  →1 →4 →3 →2  5  6  7  10  9  8  11
0   1  5  8  10  3  9  4   2 11  7  6(0)
```

Key-form reversed:

```
  →11 →8 →9 →10 7  6  5  2  3  4  1
0   11  7  4   2 9  3  8 10  1  5  6(0)
```

Adding from the right:

```
    1  4  3   2  5  6  7  10←9←8←11↖
(0) 6  5  1  10  8  3  9   2 4 7 11 0

    11 8  9  10 7  6  5   2←3←4←1↖
(0)  6 7 11   2 4  9  3  10 8 5 1 0
```

As we intend to return to the starting-point, we have, in the work-form, added an extra 0.

The same work-form that was derived by adding from right to left can also be produced by adding from left to right if (1) the key-form is changed into the balance of 12, and (2) the work-form is moved one place farther to the left. Thus:

```
    11 8  9  10 7  6  5  2  3  4  1
(0)  6 5  1  10 8  3  9  2  4  7 11 0

    1  4  3   2  5  6  7 10  9  8 11
(0) 6  7 11   2  4  9  3 10  8  5  1 0
```

Common to all work-forms is this: the last entry is always 6. The reason for this is given by Dr. Ernest Zeisler in the

tion ist 12, so wie wir Stunden addieren: 12=0; 13=1; 14=2; etc.)

```
  →1 →4 →3 →2  5  6  7  10  9  8  11
0   1  5  8  10  3  9  4   2 11  7   6
```

In der unteren Reihe haben wir alle elf Nummern von dem Anfangspunkte 0 erreicht. Die Nummern in der oberen Reihe beherrschen die Bewegungen in der unteren Reihe: Von 0 nach 1 ist 1; von 1 nach 5 ist 4; von 5 nach 8 ist 3; u.s.w. In keiner der beiden Reihen wird eine Nummer wiederholt; deshalb wissen wir, dass unser Problem gelöst ist. Nennen wir die obere Reihe die Grund-Formel, die untere die Arbeits-Formel.

Jede Grund-Formel ergibt vier Arbeits-Formeln:

```
  →1 →4 →3 →2  5  6  7  10  9  8 11
0   1  5  8  10  3  9  4   2 11  7  6 (0)
```

Grund-Formel umgekehrt:

```
  →11 →8 →9 →10 7  6  5  2  3  4  1
0   11  7  4   2 9  3  8 10  1  5  6 (0)
```

Von rechts nach links addiert:

```
    1  4  3   2  5  6  7  10←9←8←11↖
(0) 6  5  1  10  8  3  9   2 4 7 11 0

    11 8  9  10 7  6  5   2←3←4←1↖
(0)  6 7 11   2 4  9  3  10 8 5 1 0
```

Da wir beabsichtigen nach dem Anfangspunkt zurückzukehren, haben wir in der Arbeits-Formel ein weiteres 0 hinzugefügt.

Bei Addition in der umgekehrten Richtung registrieren sich die Bewegungen, die in der Arbeits-Formel vorwärts gehen, als vorwärts in der Grund-Formel, wenn wir in dieser jede Zahl von 12 abziehen und sie dann eine Nummer nach rechts verschieben:

```
    11 8  9  10 7  6  5  2  3  4  1
(0)  6 5  1  10 8  3  9  2  4  7 11 0

    1  4  3   2  5  6  7 10  9  8 11
(0) 6  7 11   2  4  9  3 10  8  5  1 0
```

Allgemeingültig für alle Arbeits-Formeln ist dies: die letzte Ziffer ist immer 6. Die Erklärung dafür ist im Anhang

mathematical treatment of the problem in the Appendix. It is proved that (1) the base of addition must be an even number and (2) the last entry in the work-form must equal base divided by 2. As 6 is always farthest removed from the starting-point 0, we have proof that six half-tones—the tritone—is the largest interval; this is the diameter of the circle. And it logically follows that there is no straight ascending or descending line to 6: we move always in a circle.

von Dr. Ernest Zeisler gegeben. Es wird bewiesen, (1) dass die Basis eine gerade Nummer sein muss, und (2) dass die letzte Ziffer in der Arbeits-Formel gleich Basis mit 2 dividiert sein muss. Da 6 immer am weitesten von dem Anfangspunkte 0 ist, haben wir den Beweis dafür, dass sechs Halbtöne—der Tritonus—das grösste Intervall ist; dieses ist der Durchmesser des Kreises; und es folgt logisch, dass es keine gerade aufsteigende oder abfallende Linie nach 6 gibt. Wir bewegen uns immer in einem Zirkel.

KEY-FORMS

GRUND-FORMELN

```
1   6   2   5   8  10   9  11   7   4   3       2   6   3   4  10   8   7   1   5   9  11
1   6   3   4   2   5  11   9  10   8   7       2   6   3   5   8   9   1   7  10   4  11
1   6   3   4   7  11   9  10   8   5   2       2   6   3   8   9   1  10   7  11   4   5
1   6   3   5   2   4  11   8  10   9   7       2   6   3  10   7   1   8   9   5   4  11
1   6   3   7   4   5   2  11   8   9  10       2   6   3  11   7   4  10   8   1   9   5
1   6   3  10   8  11   2   4   5   9   7       2   6   5   3  11   7   1   8  10   4   9
1   6   3  10   9   4   5   2  11   8   7       2   6   5   3  11   7   9   4  10   8   1
1   6   3  11   5   2   4   9  10   8   7       2   6   5   4  10   1   7   8   3  11   9
1   6   3  11   8   9   2   7   4   5  10       2   6   5   4  11   3   8   7   1  10   9
1   6   4   3   2   5  11   9  10   7   8       2   6   5   4  11   7  10   1   9   8   3
1   6   4   3   8  10   7   2  11   5   9       2   6   5   8   1   9  10  11   7   4   3
1   6   4  11   5   2   3   8  10   7   9       2   6   5   9   1   8  10   4   7  11   3
1   6   7   3  10   8   5   4   2  11   9       2   6   5   9  11   8  10   1   3   4   7
1   6   7   3  11   4   2   5   8  10   9       2   6   7   4   3   1  10   8  11   9   5
1   6   7   8   5   2  11   4   3  10   9       2   6   7   8  10   4   3   1   5   9  11
1   6   7   8  10   3   4   2  11   5   9       2   6   9   4   1   5  10   3   7   8  11
1   6   7   8  10   3   5  11   2   4   9       2   6   9   4  10   8   1   7  11   3   5
1   6   7   8  10   9   4   2   5  11   3       2   6   9   5   1   4  10   8   7   3  11
1   6   7   8  10   9  11   5   2   4   3       2   6   9   8   3   7  10   1   5   4  11
1   6   7   8  11   2   5   4   9  10   3       2   6   9  10   1   7   8   3  11   4   5
1   6   7   9   4   5   2  11   8  10   3       2   6   9  11   3   8   7   1  10   4   5
1   6   7   9  10   8  11   4   2   5   3       2   6  11   3   5   8  10   7   9   4   1
1   6   8   5   2   4   3  11   7  10   9       2   6  11   3   7   8  10   4   1   5   9
1   6   8   7   4   3  11   5   2   9  10       2   6  11   4   5   1  10   7   3   8   9
1   6   8   7  10   9  11   5   2   3   4       2   6  11   4   5   9   8   1   7  10   3
1   6   9   4   2  11   5   3  10   8   7       2   6  11   4  10   7   1   8   9   5   3
1   6   9   5   2   4  11   8   7   3  10       2   6  11   8   7   3  10   5   1   4   9
1   6   9   5  11   2   4   3  10   8   7       2   6  11   9   5   1   3   4  10   8   7
1   6   9   5  11   2   7  10   8   3   4       2   6  11   9   5   1   7   8  10   4   3
1   6   9   7   3   8   5   2   4  11  10       3   6   1   4   2   7   9  11  10   8   5
1   6   9   7  10   8   3   2   5  11   4       3   6   1   7   2   4   9   8  10  11   4
1   6   9  10   3   4  11   2   5   8   7       3   6   1   7   9   2   4   5  10   8  11
1   6   9  10   7  11   3   4   2   5   8       3   6   1  10   5   4   9   2   7   8  11
1   6   9  10   8   5   2   4  11   3   7       3   6   1  10   8   7   2   4   9   5  11
1   6   9  11   2   4   5   8  10   3   7       3   6   2   5   1   8   9  10  11   7   4
1   6  10   3   7   8  11   4   2   5   9       3   6   2   8   1   5   9   7  11  10   4
1   6  10   5   4   7   2   9   8  11   3       3   6   2   8   9   1   5   4  11   7  10
1   6  10   9   2   5  11   3   4   7   8       3   6   2  11   4   5   9   1   8   7  10
1   6  10   9   8  11   2   5   4   7   3       3   6   2  11   7   8   1   5   9   4  10
1   6  10  11   4   2   5   8   3   7   9       3   6   4   1   2   7   9  11  10   5   8
2   6   1   4   9   7  10   8   5   3  11       3   6   4   1   5  10  11   7   9   2   8
2   6   1   8  10   4   9   7  11   3   5       3   6   4   7  11  10   9   8   1   5   2
2   6   3   4   7  11  10   9   1   8   5       3   6   4  10   9  11   7   2   1   5   8
```

[23]

3	6	4	10	11	7	9	5	1	8	2
3	6	5	2	1	8	9	10	11	4	7
3	6	5	2	4	11	10	8	9	1	7
3	6	5	8	10	11	9	7	2	4	1
3	6	5	11	9	10	8	1	2	4	7
3	6	5	11	10	8	9	4	2	7	1
3	6	7	1	8	10	9	2	4	5	11
3	6	7	1	9	8	10	11	4	2	5
3	6	7	4	2	1	8	10	9	11	5
3	6	7	4	11	10	9	8	1	2	5
3	6	7	10	8	1	9	5	4	2	11
3	6	8	2	7	11	9	1	5	4	10
3	6	8	2	9	7	11	10	5	1	4
3	6	8	5	1	2	7	11	9	10	4
3	6	8	5	10	11	9	7	2	1	4
3	6	8	11	7	2	9	4	5	1	10
3	6	10	1	5	4	9	2	7	11	8
3	6	10	4	5	1	9	11	7	2	8
3	6	10	4	9	5	1	8	7	11	2
3	6	10	7	8	1	9	5	4	11	2
3	6	10	7	11	4	5	1	9	8	2
3	6	11	2	4	5	9	1	8	10	7
3	6	11	5	4	2	9	10	8	1	7
3	6	11	5	9	4	2	7	8	10	1
3	6	11	8	7	2	9	4	5	10	1
3	6	11	8	10	5	4	2	9	7	1
4	6	1	3	7	11	9	2	8	10	5
4	6	1	8	2	5	11	7	9	10	3
4	6	1	8	10	3	5	2	11	7	9
4	6	1	8	10	9	7	11	5	2	3
4	6	1	9	7	10	8	5	3	2	11
4	6	1	10	5	11	7	9	2	8	3
4	6	3	1	5	2	11	9	10	8	7
4	6	3	2	5	11	7	9	10	8	1
4	6	3	8	2	9	7	11	5	10	1
4	6	3	10	9	7	11	5	2	8	1
4	6	5	2	9	11	8	10	1	3	7
4	6	5	10	8	2	9	11	7	3	1
4	6	7	3	1	10	8	11	9	2	5
4	6	7	8	2	11	5	1	3	10	9
4	6	7	8	10	3	1	5	11	2	9
4	6	7	8	10	9	11	2	5	1	3
4	6	7	9	1	5	3	2	8	10	11
4	6	7	10	11	5	1	3	2	8	9
4	6	9	2	11	5	1	3	10	8	7
4	6	9	7	11	2	5	3	10	8	1
4	6	9	8	2	3	1	5	11	10	7
4	6	9	10	3	1	5	11	2	8	7
4	6	11	2	3	5	8	10	7	9	1
4	6	11	10	8	2	3	5	1	9	7
5	6	2	1	8	11	10	9	4	7	3
5	6	2	3	11	4	7	8	10	1	9
5	6	2	7	8	10	1	4	3	11	9
5	6	2	9	4	7	10	1	8	11	3
5	6	2	9	10	1	7	3	8	11	4
5	6	3	1	10	8	7	4	2	9	11
5	6	3	2	4	7	10	8	1	9	11
5	6	3	2	9	8	1	10	7	4	11
5	6	3	7	1	10	8	9	2	4	11
5	6	3	7	4	9	10	11	8	1	2
5	6	3	8	10	1	7	9	2	4	11
5	6	3	8	11	7	9	2	4	1	10
5	6	3	11	8	1	10	7	4	9	2
5	6	4	1	10	8	3	7	11	2	9
5	6	4	11	2	9	7	1	10	3	8
5	6	4	11	8	3	7	1	10	9	2
5	6	8	3	4	2	11	10	7	1	9
5	6	8	3	10	1	7	9	2	11	4
5	6	8	7	1	10	3	4	2	11	9
5	6	9	1	7	10	8	3	2	4	11
5	6	9	1	7	10	11	2	4	3	8
5	6	9	1	10	8	7	4	11	3	2
5	6	9	2	3	8	7	10	1	4	11
5	6	9	2	4	1	10	8	7	3	11
5	6	9	2	11	7	3	8	10	1	4
5	6	9	7	10	8	1	4	2	3	11
5	6	9	8	10	7	1	3	2	4	11
5	6	9	11	2	4	3	10	1	7	8
5	6	9	11	3	4	1	10	8	7	2
5	6	10	1	4	2	9	7	11	8	3
5	6	11	3	2	4	1	8	10	7	9
5	6	11	3	7	8	10	1	4	2	9
5	6	11	4	1	10	7	8	3	2	9
5	6	11	4	2	3	1	7	10	8	9
5	6	11	4	2	3	8	10	7	1	9
5	6	11	4	2	9	7	1	10	8	3
5	6	11	4	2	9	8	10	1	7	3
5	6	11	4	7	10	1	8	9	2	3
5	6	11	9	1	8	10	7	4	2	3
5	6	11	9	2	4	7	8	10	1	3
7	6	1	3	10	8	5	4	2	11	9
7	6	1	3	11	4	2	5	8	10	9
7	6	1	8	5	2	11	4	3	10	9
7	6	1	8	10	3	4	2	11	5	9
7	6	1	8	10	3	5	11	2	4	9
7	6	1	8	10	9	4	2	5	11	3
7	6	1	8	10	9	11	5	2	4	3
7	6	1	8	11	2	5	4	9	10	3
7	6	1	9	5	4	2	11	8	10	3
7	6	1	9	10	8	11	4	2	5	3
7	6	2	11	8	10	3	5	1	4	9
7	6	3	1	9	8	11	2	4	5	10
7	6	3	1	10	8	9	2	11	5	4
7	6	3	4	2	5	11	9	10	8	1

```
7  6  3   5   2   4  11   8  10   9   1      9  6  1   7   3   8  10   5   4   2  11
7  6  3  10   1   5   9   4   2  11   8      9  6  1   7   8  10   3   2   4  11   5
7  6  3  10   8  11   2   4   5   9   1      9  6  1  10   8   7   3  11   4   2   5
7  6  3  10   9   4   5   2  11   8   1      9  6  2   5   1   8   7  11   3   4  10
7  6  3  11   2   4   5   8   1   9  10      9  6  2   5   4  11   3   7   8   1  10
7  6  3  11   5   2   1  10   8   9   4      9  6  2   8   3   7  11   4   5   1  10
7  6  3  11   5   2   4   9  10   8   1      9  6  2   8   7  11   3   1   5  10   4
7  6  4   5  11   2   9   8  10   1   3      9  6  2  11   7   8   3  10   5   1   4
7  6  4   9   2  11   5   3  10   1   8      9  6  4   1   5  10   3   8   7  11   2
7  6  4   9   8  10   1   2   5  11   3      9  6  4   7   2   1   3   5  10  11   8
7  6  8   1   4   9   5  11   2   3  10      9  6  4   7  11  10   5   1   3   2   8
7  6  8   1  10   3   5  11   2   9   4      9  6  4  10   3   5   1   2   7  11   8
7  6  8  11   2   4   9   5   1  10   3      9  6  4  10   5   1   3  11   7   8   2
7  6  9   1   4  11   2   5   8   3  10      9  6  5   2   4  11   3   7   8  10   1
7  6  9   4   1   5   3  10   8  11   2      9  6  5   8   1   2   3   4  11  10   7
7  6  9   4   2  11   5   3  10   8   1      9  6  5   8  10  11   4   2   3   1   7
7  6  9   5   8   3   2   1   4  11  10      9  6  5  11   3   4   2   1   8  10   7
7  6  9   5  11   2   4   3  10   8   1      9  6  5  11   4   2   3  10   8   7   1
7  6  9  10   3   4  11   2   5   8   1      9  6  7   1   2   4   3   8  10   5  11
7  6  9  10   8   5   2   4  11   3   1      9  6  7   1   3   2   4  11  10   8   5
7  6  9  11   2   4   5   8  10   3   1      9  6  7   4   2   1   3   5  10   8  11
7  6 10   3   2  11   5   9   4   1   8      9  6  7  10   8   1   2   4   3  11   5
7  6 10   3   8   5   2  11   4   1   9      9  6  7  10  11   4   3   2   1   8   5
7  6 10   5   4   2  11   8   9   1   3      9  6  8   2   1   5   3   7  11   4  10
7  6 10   9   1   8   5   4   2  11   3      9  6  8   2   3   1   5  10  11   7   4
7  6 10  11   4   1   2   3   8   5   9      9  6  8   5   1   2   3   4  11   7  10
8  6  1   2   4  10   9   7  11   3   5      9  6  8  11   7   2   1   5   3  10   4
8  6  1  10   9   7   4   2   5   3  11      9  6  8  11  10   5   3   1   2   7   4
8  6  3   2   9  11   7   1  10   4   5      9  6 10   1   5   4  11   7   3   8   2
8  6  3   4  10   9  11   7   1   2   5      9  6 10   1   8   7   3  11   4   5   2
8  6  3   5   1  10   7   9   2   4  11      9  6 10   4   3  11   7   8   1   5   2
8 ·6  3  10   1   7  11   9   2   4   5      9  6 10   4  11   7   3   5   1   2   8
8  6  5   2   1   7  11   9  10   4   3      9  6 10   7  11   4   3   2   1   5   8
8  6  5   3  11   7   9  10   4   2   1      9  6 11   2   4   5  10   8   3   7   1
8  6  5   4   2   3   1  10   7  11   9      9  6 11   2   7   8   3  10   5   4   1
8  6  5   4   2   9  11   7   1  10   3      9  6 11   5   3  10   8   7   2   4   1
8  6  5   4  10   1   7  11   9   2   3      9  6 11   5  10   8   3   4   2   1   7
8  6  5   9  11   2   4   1   3  10   7      9  6 11   8  10   5   3   1   2   4   7
8  6  7   2   4  10   3   1   5   9  11     10  6  1   3   7  11   5   4   2   8   9
8  6  7  10   3   1   4   2  11   9   5     10  6  1   3   7  11   9   8   2   4   5
8  6  9   2   3   5   1   7  10   4  11     10  6  1   4   5   9   2   7  11   8   3
8  6  9   4  10   3   5   1   7   2  11     10  6  1   8   2   5  11   4   3   7   9
8  6  9  10   7   1   5   3   2   4  11     10  6  1   8   7   3   4  11   5   2   9
8  6  9  11   7  10   1   3   2   4   5     10  6  1   8   7  11   2   5   9   4   3
8  6 11   2   7   1   5   3  10   4   9     10  6  1   9   5   4   2   8  11   7   3
8  6 11   3   5   2   4   7   9  10   1     10  6  1   9   7   4   2   5   3   8  11
8  6 11   4   2   3   5   1   7  10   9     10  6  3   1   9   4   5  11   2   8   7
8  6 11   4   2   9   7  10   1   5   3     10  6  3   2  11   5   4   9   1   8   7
8  6 11   4  10   7   1   5   3   2   9     10  6  3   4   9   5   2  11   7   8   1
8  6 11   9   5   1   3  10   4   2   7     10  6  3   7  11   8   2   4   5   9   1
9  6  1   4   2   7   8  10   3   5  11     10  6  3   8   2   4  11   5   1   9   7
9  6  1   4   5  10   3   8   7   2  11     10  6  3   8  11   7   2   9   5   4   1
```

10	6	5	4	2	8	9	11	7	3	1
10	6	5	8	9	11	2	4	1	3	7
10	6	7	3	1	4	2	11	9	8	5
10	6	7	3	11	4	2	8	5	1	9
10	6	7	4	11	3	2	1	5	8	9
10	6	7	8	1	5	2	11	3	4	9
10	6	7	8	1	9	4	5	11	2	3
10	6	7	8	2	11	5	4	9	1	3
10	6	7	9	1	5	3	8	2	4	11
10	6	7	9	1	5	11	4	2	8	3
10	6	9	1	5	8	2	4	11	3	7
10	6	9	2	5	11	4	3	7	8	1
10	6	9	4	3	11	2	5	1	8	7
10	6	9	7	3	4	11	5	2	8	1
10	6	9	8	2	4	5	11	7	3	1
10	6	9	8	5	1	2	3	11	4	7
10	6	11	4	2	8	3	5	1	9	7
10	6	11	8	3	5	2	4	7	9	1
11	6	2	1	8	10	7	4	9	5	3
11	6	2	3	4	1	10	7	8	5	9
11	6	2	3	10	7	1	9	8	5	4
11	6	2	7	8	5	10	3	4	1	9
11	6	2	9	5	4	1	8	10	7	3
11	6	3	1	10	8	7	4	2	9	5
11	6	3	2	4	7	10	8	1	9	5
11	6	3	2	5	1	9	8	10	7	4
11	6	3	2	9	8	1	10	7	4	5
11	6	3	5	2	4	9	10	7	1	8
11	6	3	5	9	4	7	10	8	1	2
11	6	3	7	1	10	5	2	4	9	8
11	6	3	7	1	10	8	9	2	4	5
11	6	3	7	10	8	1	4	5	9	2
11	6	3	8	10	1	7	9	2	4	5
11	6	4	5	2	3	1	7	10	9	8
11	6	4	5	8	9	1	7	10	3	2
11	6	4	7	10	8	9	1	5	2	3
11	6	5	3	2	4	1	8	10	7	9
11	6	5	3	7	8	10	1	4	2	9
11	6	5	4	1	10	7	8	3	2	9
11	6	5	4	2	3	1	7	10	8	9
11	6	5	4	2	3	8	10	7	1	9
11	6	5	4	2	9	7	1	10	8	3
11	6	5	4	2	9	8	10	1	7	3
11	6	5	4	7	10	1	8	9	2	3
11	6	5	9	1	8	10	7	4	2	3
11	6	5	9	2	4	7	8	10	1	3
11	6	8	1	7	10	9	4	2	5	3
11	6	8	9	4	2	5	10	1	7	3
11	6	8	9	10	7	1	3	2	5	4
11	6	9	1	4	3	10	5	8	7	2
11	6	9	1	7	10	8	3	2	4	5
11	6	9	2	3	8	7	10	1	4	5

11	6	9	2	4	1	10	8	7	3	5
11	6	9	5	8	7	10	1	4	3	2
11	6	9	7	10	8	1	4	2	3	5
11	6	9	8	5	1	3	2	4	7	10
11	6	9	8	10	7	1	3	2	4	5
11	6	10	7	4	2	3	1	5	8	9
1	2	6	5	8	7	11	3	4	9	10
1	2	6	5	8	9	10	11	4	3	7
1	2	6	5	8	10	9	11	3	4	7
1	2	6	5	9	8	3	7	11	4	10
1	2	6	8	5	10	11	9	7	3	4
1	2	6	11	8	3	10	5	4	9	7
1	2	6	11	8	10	3	5	9	4	7
1	3	6	4	5	2	11	9	10	8	7
1	3	6	5	8	10	11	9	2	7	4
1	3	6	5	11	7	8	2	4	9	10
1	3	6	7	10	8	9	11	2	5	4
1	3	6	10	7	2	4	5	9	8	11
1	3	6	10	11	8	2	4	5	9	7
1	3	6	11	8	2	7	9	4	5	10
1	4	6	3	7	10	9	11	5	2	8
1	4	6	3	8	9	2	7	11	5	10
1	4	6	5	10	7	11	2	9	8	3
1	4	6	5	10	8	9	2	11	7	3
1	4	6	8	3	10	7	11	2	5	9
1	4	6	8	9	10	7	11	2	5	3
1	4	6	9	2	11	7	10	5	8	3
1	4	6	9	7	11	2	5	10	3	8
1	4	6	9	11	2	7	10	8	5	3
1	7	6	3	10	4	9	5	2	11	8
1	7	6	8	5	2	11	3	4	10	9
1	7	6	8	9	10	4	2	5	11	3
1	7	6	8	9	10	11	5	2	4	3
1	7	6	9	4	2	11	5	10	3	8
1	7	6	9	5	11	2	4	10	3	8
1	8	6	2	5	10	11	9	7	3	4
1	8	6	5	2	9	10	11	7	3	4
1	8	6	5	3	11	7	2	9	10	4
1	9	6	4	7	8	10	5	3	2	11
1	9	6	4	11	2	5	3	10	8	7
1	9	6	5	2	8	10	3	7	11	4
1	9	6	7	3	5	8	2	4	11	10
1	9	6	7	4	11	5	2	8	3	10
1	9	6	10	5	8	2	4	11	3	7
1	9	6	11	5	3	10	8	2	7	4
1	10	6	3	8	11	7	4	5	2	9
1	10	6	3	8	11	7	9	2	5	4
1	10	6	3	11	8	7	4	2	5	9
1	10	6	9	2	5	11	7	4	3	8
1	10	6	11	3	8	5	2	4	7	9
2	1	6	4	3	7	9	11	10	5	8

2	1	6	7	3	4	11	10	9	8	5
2	1	6	7	4	3	11	9	10	8	5
2	1	6	7	4	9	5	3	10	8	11
2	1	6	7	9	4	5	10	3	8	11
2	1	6	10	4	11	7	3	8	9	5
2	1	6	10	9	4	3	11	7	8	5
2	3	6	4	7	11	10	1	8	9	5
2	3	6	4	10	7	8	5	1	9	11
2	3	6	4	10	7	11	9	5	1	8
2	3	6	5	4	7	10	8	1	9	11
2	3	6	5	9	8	1	10	7	4	11
2	3	6	8	9	11	7	10	1	4	5
2	3	6	10	7	11	4	1	5	9	8
2	3	6	11	9	8	1	5	4	7	10
2	5	6	4	3	8	11	7	1	10	9
2	5	6	8	1	10	9	11	7	4	3
2	5	6	9	1	10	8	11	4	7	3
2	5	6	9	10	1	8	11	7	4	3
2	5	6	9	11	8	10	1	4	3	7
2	7	6	1	3	4	11	10	9	8	5
2	7	6	1	4	3	11	9	10	8	5
2	7	6	1	4	9	5	3	10	8	11
2	7	6	1	9	4	5	10	3	8	11
2	7	6	4	9	1	3	5	10	11	8
2	7	6	10	3	4	9	5	1	8	11
2	7	6	10	4	5	1	9	8	3	11
2	9	6	4	1	5	10	7	8	3	11
2	9	6	4	10	1	5	3	11	7	8
2	9	6	4	10	1	8	11	7	3	5
2	9	6	5	3	8	7	11	4	1	10
2	9	6	8	3	5	1	10	7	4	11
2	9	6	10	1	5	4	7	11	3	8
2	9	6	11	3	8	7	10	1	4	5
2	9	6	11	4	1	10	8	7	3	5
2	11	6	3	5	8	10	7	4	9	1
2	11	6	3	7	10	8	5	4	1	9
2	11	6	3	10	7	8	5	1	4	9
2	11	6	4	9	8	5	1	7	10	3
2	11	6	8	7	10	3	5	1	4	9
3	1	6	4	5	2	11	9	8	10	7
3	1	6	4	7	2	9	11	10	8	5
3	1	6	7	8	10	11	9	2	5	4
3	1	6	7	9	5	4	2	8	11	10
3	1	6	10	5	4	9	7	2	8	11
3	1	6	10	9	4	2	8	7	11	5
3	1	6	11	8	9	5	4	2	7	10
3	2	6	5	4	1	10	7	11	9	8
3	2	6	5	9	8	1	10	11	7	4
3	2	6	8	1	5	9	11	7	10	4
3	2	6	8	9	5	1	4	11	7	10
3	2	6	10	7	4	5	1	8	9	11
3	2	6	11	4	7	10	1	8	9	5
3	2	6	11	9	1	5	8	7	10	4
3	2	6	11	9	1	8	10	7	4	5
3	4	6	1	2	7	10	11	9	5	8
3	4	6	1	8	7	11	5	2	9	10
3	4	6	1	8	10	9	11	5	2	7
3	4	6	8	5	2	7	11	10	9	1
3	4	6	10	5	1	9	7	11	2	8
3	4	6	10	9	1	5	8	7	11	2
3	5	6	2	1	8	9	11	10	4	7
3	5	6	2	9	8	10	4	11	7	1
3	5	6	7	4	9	1	8	10	11	2
3	5	6	8	1	10	7	9	4	2	11
3	5	6	8	11	10	9	7	2	4	1
3	5	6	11	4	2	9	7	10	1	8
3	5	6	11	9	1	8	10	4	7	2
3	7	6	1	8	10	9	11	2	5	4
3	7	6	1	9	5	4	2	8	11	10
3	7	6	4	1	2	8	10	9	11	5
3	7	6	4	5	1	9	10	8	2	11
3	7	6	5	2	9	11	10	8	1	4
3	7	6	10	5	4	2	8	11	9	1
3	7	6	10	9	8	2	11	5	4	1
3	8	6	2	1	5	9	11	7	10	4
3	8	6	2	9	5	1	4	11	7	10
3	8	6	4	1	10	11	7	2	9	5
3	8	6	5	4	2	9	7	1	10	11
3	8	6	5	4	11	7	1	10	9	2
3	8	6	5	10	11	2	7	9	1	4
3	10	6	1	8	5	2	11	7	9	4
3	10	6	1	9	4	5	2	7	11	8
3	10	6	2	11	8	1	5	4	9	7
3	10	6	4	5	1	9	7	11	2	8
3	10	6	4	9	1	5	8	7	11	2
3	10	6	7	8	11	2	5	4	9	1
3	10	6	7	9	5	1	4	11	2	8
3	10	6	7	9	5	4	2	11	8	1
3	11	6	1	10	9	7	2	4	5	8
3	11	6	2	1	8	10	4	7	9	5
3	11	6	2	9	4	10	7	1	8	5
3	11	6	5	4	2	9	7	10	1	8
3	11	6	5	9	1	8	10	4	7	2
3	11	6	8	1	5	9	2	4	10	7
3	11	6	8	5	10	4	2	9	7	1
4	1	6	3	5	2	11	7	10	9	8
4	1	6	3	5	8	10	7	2	11	9
4	1	6	3	7	11	2	9	8	10	5
4	1	6	3	8	5	10	7	11	2	9
4	1	6	3	8	9	2	11	7	10	5
4	1	6	8	2	5	11	9	10	7	3
4	1	6	8	3	10	5	2	11	7	9
4	1	6	9	5	2	11	7	10	3	8
4	1	6	10	5	11	7	2	9	8	3

4	3	6	1	9	10	11	7	2	5	8	5	9	6	2	1	4	10	8	7	3	11
4	3	6	2	11	7	8	5	1	9	10	5	9	6	7	1	3	2	4	10	11	8
4	3	6	7	2	5	11	9	10	8	1	5	9	6	8	7	10	1	3	2	4	11
4	3	6	8	2	11	7	9	1	5	10	5	9	6	8	11	4	2	1	3	10	7
4	3	6	8	5	9	11	10	7	2	1	5	9	6	11	3	1	4	10	8	7	2
4	3	6	10	9	2	5	11	7	8	1	5	9	6	11	8	7	1	10	4	3	2
4	5	6	8	2	9	10	11	7	3	1	5	10	6	1	4	2	9	7	3	8	11
4	5	6	8	3	11	7	2	9	10	1	5	10	6	1	4	9	2	7	8	3	11
4	5	6	8	3	11	9	7	2	1	10	5	10	6	1	4	11	7	3	8	9	2
4	7	6	3	11	2	5	1	10	9	8	5	10	6	1	9	4	3	11	7	8	2
4	7	6	8	2	11	5	3	10	1	9	5	10	6	4	1	2	7	9	11	3	8
4	7	6	8	9	10	11	2	5	1	3	5	10	6	7	4	2	3	1	9	8	11
4	7	6	9	1	5	2	3	8	10	11	5	10	6	7	4	3	2	1	8	9	11
4	7	6	9	8	3	2	5	1	10	11	5	11	6	3	2	8	9	1	10	7	4
4	7	6	9	8	11	10	1	5	2	3	5	11	6	4	1	10	7	3	8	2	9
4	7	6	9	11	2	5	1	10	3	8	5	11	6	4	9	2	7	1	10	8	3
4	7	6	9	11	8	10	1	2	5	3	5	11	6	4	9	2	8	10	1	7	3
4	7	6	10	11	5	1	2	3	8	9	5	11	6	9	1	7	10	8	2	3	4
4	9	6	1	2	11	5	3	10	8	7	5	11	6	9	8	10	7	1	2	3	4
4	9	6	2	5	1	8	11	7	3	10	7	1	6	3	4	2	5	11	10	9	8
4	9	6	7	3	10	5	1	2	11	8	7	1	6	3	11	5	2	4	10	9	8
4	9	6	8	2	5	1	3	7	11	10	7	1	6	8	3	10	4	2	11	5	9
4	9	6	8	11	3	5	10	1	2	7	7	1	6	8	3	10	5	11	2	4	9
4	9	6	10	3	2	11	5	1	8	7	7	1	6	8	11	2	5	9	4	10	3
4	11	6	8	2	3	10	5	1	9	7	7	1	6	9	10	4	3	11	2	5	8
4	11	6	8	9	5	1	2	3	10	7	7	2	6	5	8	9	10	11	4	3	1
4	11	6	8	9	5	3	1	2	7	10	7	2	6	5	8	10	9	11	3	4	1
5	2	6	3	4	7	11	8	1	10	9	7	2	6	8	11	10	5	3	1	9	4
5	2	6	3	4	7	11	9	10	1	8	7	2	6	11	3	8	9	1	5	4	10
5	2	6	3	7	4	11	8	10	1	9	7	2	6	11	8	1	5	9	4	3	10
5	2	6	7	3	4	1	10	8	11	9	7	2	6	11	8	3	10	5	4	9	1
5	2	6	9	10	1	7	11	8	3	4	7	2	6	11	8	10	3	5	9	4	1
5	3	6	1	4	2	7	9	10	11	8	7	3	6	1	4	5	11	2	8	9	10
5	3	6	1	7	11	4	10	8	9	2	7	3	6	1	9	11	8	2	4	5	10
5	3	6	2	7	4	10	8	1	9	11	7	3	6	4	1	8	10	11	9	2	5
5	3	6	2	11	10	8	1	9	4	7	7	3	6	4	5	2	11	9	10	8	1
5	3	6	7	4	10	11	9	8	1	2	7	3	6	5	11	9	10	8	2	1	4
5	3	6	8	1	10	7	9	2	4	11	7	3	6	10	11	8	2	4	5	9	1
5	3	6	11	2	4	9	7	10	1	8	7	3	6	11	2	8	10	9	1	5	4
5	4	6	1	3	7	11	10	9	2	8	7	4	6	3	1	5	2	11	10	9	8
5	4	6	1	10	9	2	7	11	3	8	7	4	6	3	2	5	1	10	11	8	9
5	4	6	10	1	2	7	9	11	3	8	7	4	6	3	5	2	1	10	8	11	9
5	8	6	1	2	4	9	10	7	11	3	7	4	6	8	3	10	1	5	2	11	9
5	8	6	1	2	11	7	10	9	4	3	7	4	6	8	9	10	1	5	2	11	9
5	8	6	3	4	9	10	11	7	1	2	7	4	6	9	1	10	3	5	11	2	8
5	8	6	3	11	2	9	7	1	10	4	7	4	6	9	8	3	2	1	5	11	10
5	8	6	4	3	2	11	7	10	1	9	7	4	6	11	10	1	5	2	3	8	9
5	8	6	4	9	2	11	7	10	1	3	7	4	6	11	10	8	3	2	5	1	9
5	8	6	9	7	10	11	2	4	1	3	7	8	6	2	11	10	5	3	1	9	4
5	8	6	9	10	7	11	2	1	4	3	7	8	6	11	2	3	10	5	1	9	4
5	8	6	9	11	7	10	1	2	3	4	7	8	6	11	9	5	1	2	3	10	4
5	9	6	1	10	4	2	3	11	7	8	7	9	6	1	10	8	3	5	2	11	4

7	9	6	4	11	2	5	3	10	8	1	9	1	6	10	11	4	2	8	5	3	7
7	9	6	5	8	2	1	3	4	11	10	9	1	6	11	2	3	5	10	8	7	4
7	9	6	10	1	2	4	11	3	8	5	9	2	6	5	3	7	8	10	1	4	11
7	9	6	10	5	8	2	4	11	3	1	9	2	6	5	3	7	11	8	1	10	4
7	9	6	11	5	1	8	2	4	3	10	9	2	6	5	4	1	10	7	8	3	11
7	9	6	11	8	10	5	3	2	1	4	9	2	6	8	3	11	7	4	5	1	10
7	10	6	3	2	11	5	1	4	9	8	9	2	6	8	7	11	3	5	1	10	4
7	10	6	5	9	8	11	2	4	1	3	9	2	6	10	1	4	11	7	8	3	5
7	10	6	9	5	8	1	4	2	11	3	9	2	6	11	3	8	7	10	5	1	4
7	10	6	9	8	5	1	3	2	11	4	9	2	6	11	4	7	10	1	5	3	8
7	10	6	9	8	5	1	4	11	2	3	9	4	6	7	2	1	10	5	3	11	8
8	1	6	4	3	7	9	11	10	5	2	9	4	6	7	8	1	5	11	2	3	10
8	1	6	4	3	7	11	10	9	2	5	9	4	6	7	8	10	3	5	11	2	1
8	1	6	4	10	9	2	7	11	3	5	9	4	6	8	11	2	1	5	10	3	7
8	3	6	2	9	10	1	7	11	4	5	9	4	6	10	3	7	11	8	1	5	2
8	3	6	4	1	9	7	2	11	10	5	9	4	6	10	11	7	3	1	5	2	8
8	3	6	4	10	7	11	9	5	1	2	9	5	6	2	3	4	10	1	7	8	11
8	3	6	5	9	2	7	11	10	1	4	9	5	6	2	7	8	10	4	1	3	11
8	3	6	10	7	11	4	1	5	9	2	9	5	6	7	10	3	1	2	4	11	8
8	3	6	11	10	1	7	9	2	4	5	9	5	6	8	7	11	3	2	4	10	1
8	5	6	2	1	7	11	10	9	4	3	9	5	6	8	11	10	4	2	3	1	7
8	5	6	3	1	4	2	11	10	7	9	9	5	6	11	3	7	8	10	4	1	2
8	5	6	3	1	10	7	11	2	9	4	9	5	6	11	4	2	3	1	10	7	8
8	5	6	3	4	1	2	11	7	10	9	9	7	6	1	3	11	4	2	8	5	10
8	5	6	3	4	9	10	7	11	2	1	9	7	6	1	8	10	3	5	2	11	4
8	5	6	3	11	7	10	9	4	2	1	9	7	6	4	1	2	3	5	10	8	11
8	5	6	4	3	2	1	10	7	11	9	9	7	6	4	11	2	5	3	8	10	1
8	5	6	4	10	1	7	9	2	11	3	9	7	6	5	8	3	11	4	2	1	10
8	5	6	9	1	10	7	11	2	3	4	9	7	6	10	3	4	2	8	1	5	11
8	7	6	4	9	1	3	5	10	11	2	9	7	6	10	11	4	3	1	2	8	5
8	7	6	4	9	1	5	10	3	2	11	9	8	6	2	3	11	7	4	5	1	10
8	7	6	4	10	3	2	1	5	9	11	9	8	6	2	7	11	3	5	1	10	4
8	9	6	2	3	10	7	1	5	4	11	9	8	6	4	7	10	5	1	2	3	11
8	9	6	4	7	3	1	2	5	10	11	9	8	6	11	4	2	3	1	7	10	5
8	9	6	4	10	1	5	3	11	7	2	9	8	6	11	4	5	1	7	10	3	2
8	9	6	5	10	7	1	3	2	4	11	9	8	6	11	10	5	2	1	3	7	4
8	9	6	10	1	5	4	7	11	3	2	9	10	6	1	3	11	4	2	5	8	7
8	9	6	11	3	2	1	5	10	7	4	9	10	6	1	3	11	7	4	5	2	8
8	11	6	2	7	1	5	10	3	4	9	9	10	6	1	8	5	2	11	4	3	7
8	11	6	3	7	10	1	5	2	9	4	9	10	6	2	5	8	7	11	4	3	1
8	11	6	4	9	2	7	10	1	5	3	9	10	6	4	3	7	11	8	1	5	2
8	11	6	4	10	7	1	3	2	5	9	9	10	6	4	11	7	3	1	5	2	8
8	11	6	9	4	3	10	1	5	2	7	9	10	6	7	3	4	11	2	1	5	8
8	11	6	9	4	7	2	5	1	10	3	9	10	6	7	8	11	2	5	1	3	4
8	11	6	9	5	1	10	3	4	2	7	9	11	6	1	4	3	7	8	10	5	2
8	11	6	9	7	4	2	5	10	1	3	9	11	6	2	3	8	10	4	5	1	7
8	11	6	9	7	10	1	5	2	3	4	9	11	6	2	7	8	3	5	10	4	1
9	1	6	4	7	2	8	10	3	5	11	9	11	6	5	3	7	8	10	4	1	2
9	1	6	4	11	7	3	10	8	2	5	9	11	6	5	4	2	3	1	10	7	8
9	1	6	7	3	11	4	2	8	5	10	9	11	6	8	5	10	3	1	2	4	7
9	1	6	7	8	10	3	5	2	11	4	9	11	6	8	7	10	1	3	4	2	5
9	1	6	10	3	8	2	5	11	4	7	10	1	6	4	5	2	9	7	11	8	3

10	1	6	8	3	4	7	11	5	2	9
10	1	6	9	2	5	4	7	11	8	3
10	1	6	9	5	2	4	7	8	11	3
10	1	6	9	7	4	2	5	8	3	11
10	3	6	1	8	11	2	4	5	9	7
10	3	6	1	9	4	5	2	11	8	7
10	3	6	2	11	7	8	5	1	9	4
10	3	6	4	9	7	11	2	5	8	1
10	3	6	7	9	4	5	1	8	11	2
10	3	6	8	2	11	4	1	5	9	7
10	3	6	8	2	11	7	9	1	5	4
10	3	6	8	11	7	2	5	4	9	1
10	5	6	2	8	7	11	3	4	9	1
10	5	6	2	9	8	3	7	11	4	1
10	5	6	8	3	11	9	7	2	1	4
10	5	6	11	3	8	7	2	9	4	1
10	5	6	11	8	3	7	9	2	4	1
10	5	6	11	8	9	1	3	2	4	7
10	5	6	11	9	8	1	2	3	4	7
10	7	6	3	1	4	2	11	8	9	5
10	7	6	3	2	11	4	1	5	8	9
10	7	6	3	11	2	4	1	8	5	9
10	7	6	4	11	2	3	1	5	8	9
10	7	6	8	9	4	1	5	11	2	3
10	9	6	1	3	4	11	7	8	5	2
10	9	6	2	5	1	8	11	7	3	4
10	9	6	4	3	1	5	2	11	8	7
10	9	6	7	3	4	11	2	5	8	1
10	9	6	7	8	5	2	4	11	3	1
10	9	6	8	2	5	1	3	7	11	4
10	9	6	8	2	5	4	7	11	3	1
10	9	6	8	5	1	2	11	4	3	7
10	11	6	2	3	8	9	1	5	4	7
10	11	6	2	8	1	5	9	4	3	7
10	11	6	5	3	8	7	2	9	4	1
10	11	6	5	8	3	7	9	2	4	1
10	11	6	5	8	9	1	3	2	4	7
10	11	6	5	9	8	1	2	3	4	7
10	11	6	8	9	5	3	1	2	7	4
11	2	6	1	9	4	7	10	8	5	3
11	2	6	3	10	7	1	5	8	9	4
11	2	6	9	1	4	5	8	10	7	3
11	2	6	9	4	1	5	3	10	7	8
11	2	6	9	4	1	5	8	7	10	3
11	3	6	1	7	9	2	4	10	5	8
11	3	6	2	7	4	10	8	1	9	5
11	3	6	5	8	1	7	10	4	9	2
11	3	6	5	9	7	4	10	8	1	2
11	3	6	7	10	4	2	9	5	1	8
11	3	6	8	1	10	7	9	2	4	5
11	3	6	8	5	4	2	7	9	10	1
11	4	6	7	9	1	5	10	3	2	8

11	4	6	7	10	3	2	1	5	9	8
11	4	6	10	7	2	1	3	5	9	8
11	5	6	3	7	1	10	8	2	9	4
11	5	6	3	8	10	1	7	2	9	4
11	5	6	4	3	2	1	7	10	8	9
11	5	6	4	3	2	8	10	7	1	9
11	5	6	4	7	10	1	9	8	2	3
11	5	6	9	2	8	3	7	10	1	4
11	8	6	3	1	10	5	2	4	7	9
11	8	6	3	5	1	10	7	2	9	4
11	8	6	3	10	1	5	2	7	4	9
11	8	6	4	3	2	5	1	10	7	9
11	8	6	4	9	2	5	1	10	7	3
11	8	6	7	2	4	3	10	1	5	9
11	8	6	7	2	5	1	10	3	4	9
11	8	6	9	4	3	10	5	1	7	2
11	8	6	9	5	2	3	1	7	10	4
11	9	6	1	4	10	5	3	8	7	2
11	9	6	2	1	4	10	8	7	3	5
11	9	6	2	5	10	8	7	3	4	1
11	9	6	5	2	4	3	1	10	7	8
11	9	6	7	1	5	4	10	8	3	2
11	9	6	7	4	2	1	3	10	5	8
11	9	6	8	7	10	1	3	2	4	5
11	10	6	1	4	2	9	7	3	8	5
11	10	6	1	4	9	2	7	8	3	5
11	10	6	4	7	2	1	3	5	9	8
11	10	6	7	3	4	9	5	1	8	2
11	10	6	7	4	2	3	1	9	8	5
11	10	6	7	4	3	2	1	8	9	5
11	10	6	7	4	5	1	9	8	3	2
1	2	5	6	3	4	10	9	7	11	8
1	2	5	6	8	7	11	3	4	10	9
1	2	7	6	4	3	10	8	9	5	11
1	2	7	6	5	11	3	8	10	9	4
1	2	8	6	9	5	3	11	7	4	10
1	3	5	6	2	9	8	10	11	4	7
1	3	5	6	8	11	10	9	2	7	4
1	3	5	6	11	8	7	2	4	9	10
1	3	5	6	11	8	10	9	2	4	7
1	3	7	6	9	8	5	4	2	11	10
1	3	11	6	5	8	7	2	4	9	10
1	3	11	6	5	8	10	9	2	4	7
1	4	5	6	3	8	11	7	2	9	10
1	4	5	6	10	7	2	9	11	8	3
1	4	5	6	10	9	8	2	11	7	3
1	4	9	6	2	5	8	10	7	3	11
1	4	9	6	11	3	5	8	10	7	2
1	4	10	6	5	2	7	9	11	3	8
1	4	11	6	10	7	8	3	5	2	9
1	7	2	6	11	8	3	5	10	4	9

1	7	3	6	11	10	8	5	4	2	9
1	7	8	6	5	11	3	2	4	10	9
1	7	9	6	4	11	2	5	10	3	8
1	7	9	6	10	5	2	11	4	3	8
1	7	9	6	11	4	2	5	10	8	3
1	8	5	6	2	9	10	11	7	4	3
1	8	5	6	3	4	7	9	10	11	2
1	8	5	6	3	11	7	10	4	9	2
1	8	7	6	5	11	3	2	4	9	10
1	8	7	6	10	3	4	2	9	5	11
1	8	11	6	3	10	7	9	4	5	2
1	9	5	6	2	8	7	3	11	4	10
1	9	5	6	11	8	10	3	2	4	7
1	9	7	6	10	5	2	4	11	8	3
1	9	10	6	5	2	8	11	7	4	3
1	9	10	6	7	8	2	4	5	11	3
1	9	11	6	2	3	8	10	5	4	7
1	9	11	6	5	8	10	3	2	4	7
1	10	4	6	5	8	7	3	11	9	2
1	10	5	6	4	7	11	9	2	8	3
1	10	5	6	9	2	8	3	7	11	4
1	10	9	6	8	5	2	4	7	3	11
2	1	4	6	3	7	10	11	9	5	8
2	1	4	6	10	11	7	3	8	5	9
2	1	7	6	3	4	10	11	9	8	5
2	1	10	6	4	11	7	3	8	5	9
2	3	8	6	4	5	11	7	10	1	9
2	3	8	6	4	11	10	7	1	5	9
2	3	10	6	7	4	5	9	1	8	11
2	3	11	6	9	1	7	8	10	4	5
2	3	11	6	9	8	5	1	4	10	7
2	5	3	6	7	4	10	8	11	9	1
2	5	3	6	11	8	10	4	7	9	1
2	5	4	6	3	1	7	11	10	9	8
2	5	4	6	8	3	11	7	10	1	9
2	5	4	6	8	9	10	1	7	11	3
2	5	4	6	11	9	8	1	10	7	3
2	5	8	6	1	3	4	11	7	9	10
2	5	8	6	4	3	7	11	10	9	1
2	5	10	6	9	1	4	3	11	7	8
2	7	1	6	9	4	10	5	3	8	11
2	7	4	6	9	1	10	5	3	11	8
2	7	4	6	10	5	1	9	8	11	3
2	7	10	6	4	5	1	9	8	11	3
2	8	3	6	4	10	7	11	5	9	1
2	8	3	6	10	11	7	4	5	1	9
2	8	5	6	4	3	7	9	11	10	1
2	8	9	6	4	10	1	5	11	3	7
2	8	9	6	10	5	1	4	11	7	3
2	8	11	6	4	9	1	3	5	10	7
2	9	5	6	3	7	1	8	10	4	11
2	9	5	6	3	8	11	7	4	10	1
2	9	8	6	4	5	10	1	7	11	3
2	9	8	6	4	11	5	1	10	7	3
2	9	10	6	1	4	11	3	7	8	5
2	11	4	6	5	3	8	7	10	1	9
2	11	4	6	8	3	10	7	1	5	9
2	11	4	6	8	9	5	1	10	7	3
2	11	4	6	9	7	1	5	10	3	8
2	11	8	6	4	9	1	5	10	3	7
2	11	8	6	7	9	4	5	1	3	10
2	11	9	6	1	4	10	8	5	3	7
2	11	9	6	5	8	10	4	1	3	7
2	11	10	6	3	7	4	9	5	1	8
3	1	4	6	5	10	8	9	11	2	7
3	1	4	6	7	2	11	9	10	8	5
3	1	7	6	8	9	4	5	2	11	10
3	1	7	6	8	9	10	11	2	5	4
3	1	7	6	9	8	10	11	2	4	5
3	1	10	6	5	9	7	4	2	8	11
3	1	10	6	5	9	11	8	2	4	7
3	2	8	6	9	7	11	10	1	5	4
3	2	11	6	1	9	5	8	10	7	4
3	4	7	6	2	1	10	8	11	9	5
3	4	7	6	5	9	1	10	8	11	2
3	4	10	6	5	9	1	7	11	2	8
3	4	10	6	9	5	1	8	11	7	2
3	5	2	6	1	9	7	4	10	8	11
3	5	2	6	1	9	11	8	10	4	7
3	5	8	6	1	2	4	9	7	10	11
3	5	8	6	11	10	7	9	2	4	1
3	5	11	6	4	9	2	7	10	1	8
3	5	11	6	4	9	8	1	10	7	2
3	5	11	6	9	4	2	7	10	8	1
3	7	1	6	9	2	4	5	8	10	11
3	7	4	6	1	2	8	10	11	9	5
3	8	2	6	1	9	5	11	7	10	4
3	8	2	6	9	1	5	4	7	11	10
3	8	11	6	1	9	5	2	4	7	10
3	8	11	6	10	5	2	4	7	9	1
3	10	4	6	9	11	7	2	5	1	8
3	10	7	6	5	9	1	4	2	11	8
3	11	5	6	9	10	8	1	4	2	7
3	11	8	6	5	10	4	2	7	9	1
4	1	2	6	8	5	9	11	10	7	3
4	1	2	6	9	5	8	3	7	11	10
4	1	3	6	5	8	10	9	11	2	7
4	1	3	6	7	2	11	9	8	10	5
4	1	8	6	2	5	9	11	10	7	3
4	1	8	6	3	5	11	7	2	9	10
4	1	8	6	7	9	10	11	2	5	3
4	1	9	6	11	3	5	10	8	2	7
4	1	10	6	2	9	11	7	8	3	5
4	1	10	6	5	9	2	7	11	3	8

```
4   3   8   6   5  11  10   9   2   7   1        5   4   7   6   3   2  11   9   8   1  10
4   3  10   6   9   1   5  11   2   7   8        5   4  11   6   1   7   3  10   8   9   2
4   3  10   6   9   5   2  11   7   1   8        5   4  11   6   2   3   8  10   9   1   7
4   5   2   6   3   7  10   1   8   9  11        5   8   1   6   2   9   4  10   7  11   3
4   5   2   6   3  11   7   1  10   9   8        5   8   1   6   2  11  10   9   7   4   3
4   5   2   6   8   9  10  11   7   1   3        5   8   1   6   3   4   7  11  10   9   2
4   5   2   6   9   1  10   7  11   3   8        5   8   2   6   1  10  11   9   7   3   4
4   5   8   6   2   9  10  11   7   1   3        5   8   7   6   2  11   4   3   1  10   9
4   5   8   6   3   1  10   7   2   9  11        5   8   9   6   7   3   1   4   2  11  10
4   5   8   6   3  11   2   7   9   1  10        5   8   9   6  10   1   4   2  11   3   7
4   5  11   6   3   2   8  10   9   1   7        5   9   1   6   7   4   2   3  10   8  11
4   7   2   6   3  11   8   9   1   5  10        5   9   1   6  10   4  11   3   7   8   2
4   7   2   6   8  11   3   5  10   1   9        5   9   2   6   1  10   4   7  11   8   3
4   7   3   6   5   9  11  10   8   2   1        5   9   2   6  11   4  10   8   1   7   3
4   7   8   6   1   3  10   5   2  11   9        5   9   7   6   1   4   2   3  10   8  11
4   7   8   6   2  11   3   5  10   1   9        5   9   7   6  10   3   4   2   1   8  11
4   7   8   6   9  11   5   1   2   3  10        5   9  11   6   2   1  10   8   7   4   3
4   7   9   6   1   2   5   3   8  10  11        5  10   1   6   3   8   2   9  11   7   4
4   7   9   6  11   8  10   3   5   2   1        5  10   1   6   4  11   7   3   8   2   9
4   7  10   6   2   3   5   1   8   9  11        5  10   4   6   9   1   3   7  11   8   2
4   7  10   6  11   3   2   1   5   9   8        5  10  11   6   1   7   3   4   2   9   8
4   9   8   6  11   5  10   3   2   1   7        5  10  11   6   8   3   2   4   9   1   7
4   9  10   6   3   7  11   5   2   1   8        5  11   3   6   7   2   4   1   8  10   9
4   9  10   6   3  11   2   5   1   7   8        5  11   4   6   1   7   3  10   8   2   9
4  10   1   6   2   9  11   3   7   8   5        5  11   9   6   2   1  10   7   8   3   4
4  10   3   6   8   1   5   2   7  11   9        5  11   9   6   7   8  10   1   2   4   3
4  10   5   6   2   8  11   7   3   1   9        5  11   9   6   8   7  10   1   2   3   4
4  10   7   6   2   3   5   9   1   8  11        5  11  10   6   7   4   3   1   2   8   9
4  10   9   6   8   7  11   2   1   5   3        7   1   2   6   5   8   9  11  10   4   3
4  10  11   6   2   8   5   1   9   7   3        7   1   3   6   4   5   2  11  10   9   8
4  11   2   6   3   7  10   1   5   9   8        7   1   3   6   5   4   2  11  10   8   9
4  11   2   6   8   3  10   5   1   7   9        7   1   3   6  10  11   2   5   4   9   8
4  11   2   6   9   1  10   7   8   3   5        7   1   8   6  11   5   9   2   4  10   3
4  11   2   6   9   5   1   7  10   3   8        7   1   9   6   5  10   8  11   4   2   3
4  11   5   6   9   2   8  10   3   7   1        7   2   1   6   4   9  10   8   3  11   5
4  11   8   6   2   3  10   5   1   7   9        7   2   1   6  11   5   9   8  10   3   4
4  11   8   6   9   5   2   1   3   7  10        7   2   8   6   3  11   9   5   1   4  10
4  11   8   6   9   7  10   1   2   3   5        7   2  11   6   8   1   5   9   4  10   3
5   2   1   6   8  11   7   9  10   4   3        7   2  11   6   9   4  10   3   1   5   8
5   2   1   6   9  10   4   3  11   7   8        7   3   1   6  10  11   2   4   5   8   9
5   2   8   6   1   4  11   3   7   9  10        7   3   5   6   2   9   8  10  11   4   1
5   2   9   6   4   1  10   8  11   3   7        7   3   5   6  11   8  10   9   2   4   1
5   3   1   6   4   7   2   9  10  11   8        7   3  10   6   1   8   2   4  11   5   9
5   3   1   6   7   4   2   9  10   8  11        7   3  10   6  11   2   8   5   1   4   9
5   3   1   6   7   4  11  10   8   9   2        7   3  11   6   2   8   1   9   5   4  10
5   3   1   6  10   9   4   2   7   8  11        7   3  11   6   5   8  10   9   2   4   1
5   3   7   6   1   4   2   9  10   8  11        7   4   3   6   2  11   8  10   1   9   5
5   3   7   6   1   4  11  10   8   9   2        7   4   3   6   5   9  11   8  10   1   2
5   3  11   6   9   4   1   8  10   7   2        7   4   5   6  10   1   8   9  11   2   3
5   4   1   6   3   7  11   2   8   9  10        7   4  10   6  11   2   1   3   5   9   8
5   4   1   6   3   8  11   9   2   7  10        7   4  11   6   9   8   5   1   2   3  10
5   4   1   6  10   9   2   7  11   8   3        7   4  11   6  10   1   2   3   5   8   9
```

```
7   4  11   6  10   3   8   2   5   1   9        8   7  10   6   9   5   2  11   4   3   1
7   8   1   6  10   9   4   2   3  11   5        8   9   2   6   3   7  10   1   5  11   4
7   8   1   6  11   5   9   2   4   3  10        8   9   2   6   3  11   7   1  10   5   4
7   8   5   6   9  10   1   3   4  11   2        8   9   4   6   7   1   2   3  10   5  11
7   8  11   6   2   3  10   5   1   4   9        8  11   2   6   7   3  10   5   1   9   4
7   8  11   6   9   4   1   3  10   5   2        8  11   2   6  10   3   1   5   4   9   7
7   8  11   6   9   5   1  10   4   3   2        8  11   3   6   1   9   7   2   4  10   5
7   9   1   6   3   8  11   4   2   5  10        8  11   4   6   5   3   2   1  10   7   9
7   9   5   6  11   8   1   2   4   3  10        8  11   4   6   9   7   1   5  10   3   2
7   9   5   6  11   8  10   3   2   4   1        8  11   4   6  10   7   3   1   2   5   9
7   9  11   6   2   3   8  10   5   4   1        8  11   9   6   5  10   1   3   4   2   7
7   9  11   6   5   8   1   2   4   3  10        8  11   9   6   7   4   2   3   1  10   5
7   9  11   6   5   8  10   3   2   4   1        8  11  10   6   3   7   4   9   5   1   2
7   9  11   6   8   5  10   3   2   1   4        8  11  10   6   4   7   3   1   2   5   9
7  10   3   6   8  11   2   4   1   9   5        9   1   4   6   7   2   8  10   5   3  11
7  10   4   6  11   8   1   9   5   3   2        9   1   7   6   3   2   4  11   8  10   5
7  10  11   6   3   2   8   9   1   5   4        9   2   5   6   7   3  11   8  10   1   4
7  10  11   6   4   1   5   3   2   8   9        9   2   8   6   3   1   5  10   7  11   4
8   1   4   6   3   5   2  11  10   9   7        9   4   1   6   2   7  10   8   5   3  11
8   1   4   6   3   7  10  11   9   5   2        9   4   1   6  11   3   7  10   8   5   2
8   1   4   6  10   9   2   7  11   5   3        9   4  10   6   3  11   7   8   5   1   2
8   1   7   6   3  10   4   2   9   5  11        9   4  10   6  11   3   7   1   5   2   8
8   1  10   6   3   7  11   5   2   9   4        9   5   8   6  11  10   4   2   1   3   7
8   1  10   6   3  11   2   5   4   9   7        9   5  11   6   3  10   8   7   4   2   1
8   1  10   6   4   9   2   7  11   5   3        9   7   1   6   3   8  10   5   2   4  11
8   1  10   6   9   5   2  11   7   3   4        9   7   1   6   8   3   4  11   2   5  10
8   2   1   6  10   4   7  11   3   5   9        9   7   1   6   8   3  10   5   2  11   4
8   2   3   6   4   5   1  10  11   7   9        9   7   4   6   1   2   5   3  10   8  11
8   2   5   6  10   9   7   3  11   4   1        9   7   4   6  11  10   8   3   5   2   1
8   2   7   6  10   4   1   5   9  11   3        9   7  10   6  11   3   1   4   2   8   5
8   2   9   6   4  11   7  10   5   1   3        9   7  10   6  11   3   5   8   2   4   1
8   2  11   6  10   3   1   9   5   4   7        9   8   2   6   3   7  11   4   1   5  10
8   3   2   6   9   1  10   7  11   5   4        9   8   2   6   7   3  11   5   1  10   4
8   3   2   6   9   5   1   7  10  11   4        9   8   5   6   7   3  11   2   4   1  10
8   3   4   6   1   7   2   9  10  11   5        9   8   5   6  10  11   2   4   1   3   7
8   5   2   6   1   9  10  11   7   3   4        9  10   1   6  11   3   7   4   2   5   8
8   5   2   6  10   9   7  11   4   3   1        9  10   4   6   3   5   1   2  11   7   8
8   5   3   6   1   4   2   9   7  10  11        9  11   2   6   7   3   1   4  10   8   5
8   5   3   6  11  10   7   9   4   2   1        9  11   2   6   7   3   5   8  10   4   1
8   5   4   6   3   1   7  11  10   9   2        9  11   5   6   3   4   2   1  10   8   7
8   5   4   6  10   1   9   7   2  11   3        9  11   5   6   4   3   2   1  10   7   8
8   5   4   6  11   9   2   7  10   1   3        9  11   5   6   4   3   8   7  10   1   2
8   5   9   6   7   3   1   2   4  10  11        9  11   8   6   5  10   1   3   2   4   7
8   5  10   6   4   1   9   7   2  11   3        9  11   8   6   7   2   4   3   1  10   5
8   5  10   6   9   1   4   3  11   7   2       10   1   2   6   9   5   8   3   7  11   4
8   7   1   6   9  10   4   2   3  11   5       10   1   3   6   7   4   2   8  11   9   5
8   7   4   6   9   1  10   5   3  11   2       10   1   3   6  11   8   2   4   7   9   5
8   7   4   6   9  11   2   5  10   3   1       10   1   4   6   5   3   8   7  11   9   2
8   7   4   6  10   3   2   1   5  11   9       10   1   4   6   8   3  11   7   2   9   5
8   7  10   6   3  11   2   5   1   9   4       10   1   8   6   3   5  11   7   2   9   4
8   7  10   6   4   3   2   1   5  11   9       10   1   8   6   4   3   7  11   2   5   9
8   7  10   6   9   1   5  11   2   3   4       10   1   8   6   4   9   2   5  11   7   3
```

```
10   1   8   6   7   9   4   5   2  11   3       11   5   3   6   2   7  10   1   8   9   4
10   3   2   6  11   8   1   9   5   4   7       11   5   3   6   8   1  10   7   2   9   4
10   3   4   6   8   1   7  11   2   5   9       11   5   4   6   7   1   9  10   8   2   3
10   3   4   6   8   7   2  11   5   1   9       11   5   9   6   1   2   4   7   8  10   3
10   3   7   6   9   4   1   5   8   2  11       11   5  10   6   1   4   9   7   2   8   3
10   3   7   6   9   5  11   4   2   8   1       11   8   1   6   2   5   4   9   7  10   3
10   4   1   6   8   3  11   9   7   2   5       11   8   2   6   7  10   5   3   1   9   4
10   4   3   6   2   7  11   8   1   5   9       11   8   3   6   1   9   7   4   2   5  10
10   4   3   6   8   2  11   7   1   9   5       11   8   3   6  10   7   4   2   5   9   1
10   4   7   6   8   9   5   3   1   2  11       11   8   7   6   2   3   4  10   1   5   9
10   4   9   6   2   1   5   8   7  11   3       11   8   7   6   2   5  10   3   1   4   9
10   4   9   6   8   2   5   1   7   3  11       11   8   7   6   9   4   1   5  10   3   2
10   5   2   6   8   7  11   3   4   1   9       11   9   1   6   7   4   2   3  10   8   5
10   5   8   6   2   7  11   3   4   1   9       11   9   1   6   7   4   5  10   8   3   2
10   5   8   6   3  11   2   7   9   1   4       11   9   5   6   3   4   7   8  10   1   2
10   5  11   6   3   8   2   7   9   4   1       11   9   7   6   1   4   2   3  10   8   5
10   7   2   6   3  11   8   9   1   5   4       11   9   7   6   1   4   5  10   8   3   2
10   7   4   6   8   9   5   1   2   3  11       11   9   7   6   4   1   2   3  10   5   8
10   7   4   6  11   9   8   1   5   3   2       11   9   7   6  10   3   4   2   1   8   5
10   7   8   6   1   3   4  11   2   5   9       11  10   4   6   3   7   9   1   5   8   2
10   7   8   6   4   3   2  11   5   1   9       11  10   5   6   7   1   9   4   2   3   8
10   7   8   6   4   9   1   5   2  11   3       11  10   5   6   8   9   2   4   3   7   1
10   7   8   6   9  11   5   1   2   3   4       11  10   7   6   4   5   1   9   8   2   3
10   7   9   6   1   4   2   8   5   3  11       11  10   7   6   9   8   2   3   5   1   4
10   7   9   6   5   8   2   4   1   3  11
10   9   1   6   3   4   7  11   8   2   5        1   2   5   3   6   4   7  10   8   9  11
10   9   1   6   3  11   5   4   2   8   7        1   2   5   9   6   8   3  11   7  10   4
10   9   2   6   5   8   7   3  11   4   1        1   2   7   4   6   3   8  10  11   5   9
10   9   4   6   8   1   2   5  11   7   3        1   2   8   5   6   9   7   3   4  11  10
10   9   4   6   8   7   1   5   2  11   3        1   2   8   5   6  10  11   7   3   4   9
10  11   2   6   8   1   5   9   4   7   3        1   3   5   2   6   9   8  10   7   4  11
10  11   5   6   9   8   2   1   3   4   7        1   3   7   4   6   5   8  10   9   2  11
10  11   8   6   2   1   5   9   4   7   3        1   3   7   9   6   8   5   2   4  10  11
10  11   8   6   9   5   2   1   3   7   4        1   3  10   7   6   5   9   2   4  11   8
11   2   3   6   4   7  10   8   5   9   1        1   3  11   5   6   8   7   2   4  10   9
11   2   7   6   3  10   4   9   5   1   8        1   3  11   5   6   8   9  10   4   2   7
11   2   7   6   8   5   1   3  10   4   9        1   4   2   9   6   5   8   3   7  11  10
11   2   8   6   7   4   5   9   1   3  10        1   4   3   2   6   5  10   8  11   9   7
11   3   1   6   7   4   2   9  10   8   5        1   4   3   7   6   2   5  10   8   9  11
11   3   1   6  10   9   4   2   7   8   5        1   4   5  10   6   7   2   8   9  11   3
11   3   2   6   5   4  10   8   7   1   9        1   4   5  10   6   9   8   2   7  11   3
11   3   2   6   7  10   4   1   5   8   9        1   4   9   2   6  11  10   8   5   3   7
11   3   5   6   2   7  10   8   1   4   9        1   4  10   5   6   2   7   8   3  11   9
11   3   7   6   1   4   2   9  10   8   5        1   4  10   5   6   9   8   3  11   7   2
11   3   7   6  10   4   5   9   1   8   2        1   7   2  11   6   8   5   3  10   9   4
11   4   1   6   9   2   5   3   8   7  10        1   7   3   4   6   5   8   9  10  11   2
11   4   5   6   2   9   8  10   3   7   1        1   7   3   5   6   9   2   8  10  11   4
11   4   5   6   7   1   9  10   8   3   2        1   7   3   5   6  11  10   8   2   9   4
11   4   7   6   9   1   5   2   8   3  10        1   7   3  11   6   5  10   8   2   9   4
11   4   7   6   9   8   5   3   2   1  10        1   7   3  11   6  10   5   8   2   4   9
11   4   7   6  10   3   2   1   5   8   9        1   7   9  10   6   5   2   3   4  11   8
11   5   3   6   1   8  10   7   2   4   9        1   8   2   5   6   9  10   3   7  11   4
```

1	8	5	3	6	4	7	10	11	9	2
1	8	5	3	6	11	10	7	4	9	2
1	8	7	10	6	3	4	2	5	9	11
1	8	7	10	6	9	2	4	11	5	3
1	8	11	3	6	10	7	4	5	9	2
1	9	4	3	6	8	2	7	11	5	10
1	9	4	7	6	5	3	8	10	11	2
1	9	5	2	6	10	11	8	3	7	4
1	9	5	8	6	3	11	2	7	10	4
1	9	5	11	6	3	8	10	4	7	2
1	9	5	11	6	8	3	10	4	2	7
1	9	7	10	6	5	2	4	3	8	11
1	9	11	2	6	10	5	8	3	7	4
1	10	4	5	6	8	7	11	3	2	9
1	10	5	4	6	3	2	8	7	11	9
1	10	5	11	6	8	3	2	9	7	4
1	10	8	3	6	5	11	7	2	9	4
1	10	9	7	6	8	5	4	2	3	11
2	1	4	3	6	5	11	9	8	10	7
2	1	4	3	6	7	10	11	9	8	5
2	1	4	9	6	11	8	3	5	10	7
2	1	4	10	6	11	3	7	8	5	9
2	1	8	5	6	3	7	9	4	10	11
2	1	8	5	6	3	7	11	10	4	9
2	1	10	9	6	5	8	3	11	4	7
2	3	4	1	6	7	9	11	8	10	5
2	3	4	7	6	1	9	11	8	10	5
2	3	5	11	6	4	1	8	9	10	7
2	3	8	9	6	7	4	5	1	10	11
2	3	8	9	6	11	5	1	10	4	7
2	5	1	3	6	4	7	11	10	9	8
2	5	1	9	6	10	4	3	11	7	8
2	5	3	1	6	11	4	7	10	8	9
2	5	3	7	6	4	1	9	8	11	10
2	5	3	11	6	1	4	9	8	10	7
2	5	4	11	6	1	3	7	10	8	9
2	5	4	11	6	9	7	1	8	10	3
2	5	9	1	6	4	7	3	8	11	10
2	5	10	4	6	1	9	7	3	11	8
2	7	4	3	6	5	8	9	11	10	1
2	7	4	9	6	1	10	5	3	8	11
2	7	4	9	6	11	5	3	8	10	1
2	7	4	10	6	5	9	1	8	11	3
2	7	8	11	6	9	1	3	4	10	5
2	7	8	11	6	9	1	5	10	4	3
2	7	10	3	6	11	8	9	5	4	1
2	8	3	4	6	9	7	1	5	10	11
2	8	5	4	6	3	7	10	11	9	1
2	8	5	10	6	9	1	4	11	3	7
2	8	9	4	6	3	1	7	11	10	5
2	8	11	4	6	9	1	10	5	3	7
2	8	11	10	6	3	7	4	5	9	1

2	9	4	1	6	7	3	5	8	10	11
2	9	4	7	6	1	3	5	8	10	11
2	9	8	3	6	1	4	11	7	10	5
2	9	8	3	6	5	11	7	10	4	1
2	9	11	5	6	4	7	8	3	10	1
2	11	3	7	6	4	1	9	8	5	10
2	11	4	5	6	3	1	7	8	10	9
2	11	4	5	6	7	9	1	10	8	3
2	11	7	3	6	10	4	9	5	1	8
2	11	7	9	6	4	1	5	10	3	8
2	11	9	1	6	4	7	3	8	5	10
2	11	9	5	6	7	4	3	8	10	1
2	11	9	7	6	5	4	1	10	8	3
2	11	10	4	6	7	3	1	9	5	8
3	1	4	5	6	10	9	7	2	11	8
3	1	4	5	6	10	9	8	11	2	7
3	1	5	2	6	8	9	10	11	7	4
3	1	5	8	6	2	9	10	11	7	4
3	1	9	7	6	5	10	4	2	11	8
3	1	10	5	6	7	9	4	2	11	8
3	2	4	7	6	10	5	1	9	8	11
3	2	5	4	6	1	7	9	10	8	11
3	2	11	10	6	5	8	1	9	4	7
3	4	1	2	6	5	8	9	11	10	7
3	4	1	2	6	7	10	8	9	11	5
3	4	2	5	6	8	9	10	11	7	1
3	4	7	2	6	1	10	11	9	8	5
3	4	9	1	6	10	5	11	7	2	8
3	4	9	10	6	1	2	11	7	8	5
3	4	9	10	6	5	8	2	11	7	1
3	5	1	4	6	10	9	2	7	11	8
3	5	1	10	6	4	9	2	7	11	8
3	5	2	1	6	11	9	8	10	7	4
3	5	8	1	6	2	9	4	7	10	11
3	5	8	1	6	2	9	11	10	7	4
3	5	9	11	6	1	2	8	10	7	4
3	7	1	8	6	4	9	2	5	11	10
3	7	4	5	6	10	9	1	8	11	2
3	7	9	4	6	8	1	2	5	11	10
3	7	10	11	6	8	2	5	1	9	4
3	7	11	2	6	8	1	5	9	4	10
3	7	11	8	6	2	1	5	9	4	10
3	8	5	10	6	1	4	9	7	2	11
3	8	5	10	6	11	2	4	9	7	1
3	8	9	2	6	1	4	10	7	11	5
3	8	9	2	6	5	10	7	11	4	1
3	8	9	5	6	2	1	7	11	10	4
3	8	10	1	6	4	9	2	7	11	5
3	8	11	10	6	5	2	7	9	4	1
3	10	1	8	6	5	11	9	2	4	7
3	10	7	2	6	1	4	5	9	8	11
3	10	8	11	6	2	1	5	9	4	7

3	11	2	7	6	4	10	1	5	9	8
3	11	5	4	6	8	9	10	1	7	2
3	11	7	4	6	10	5	1	9	8	2
3	11	7	10	6	4	5	1	9	8	2
3	11	8	1	6	2	9	5	4	7	10
3	11	9	8	6	4	5	10	1	7	2
4	1	2	7	6	3	11	5	10	8	9
4	1	2	8	6	5	9	11	3	7	10
4	1	2	8	6	5	11	9	10	3	7
4	1	5	3	6	8	11	7	2	9	10
4	1	5	9	6	2	8	3	7	11	10
4	1	9	5	6	8	2	11	10	7	3
4	1	9	7	6	5	2	3	10	8	11
4	3	1	9	6	10	5	11	2	7	8
4	3	7	1	6	2	11	10	9	8	5
4	3	8	2	6	11	10	5	1	7	9
4	5	2	3	6	11	8	10	9	7	1
4	5	2	8	6	9	10	7	11	3	1
4	5	8	2	6	1	9	11	10	7	3
4	5	10	1	6	9	11	7	8	2	3
4	5	11	3	6	2	7	1	10	9	8
4	7	2	1	6	9	5	11	10	8	3
4	7	2	8	6	11	3	5	9	1	10
4	7	2	8	6	11	5	3	10	9	1
4	7	3	1	6	11	2	9	10	8	5
4	7	3	11	6	8	2	5	10	1	9
4	7	11	3	6	2	8	9	1	5	10
4	7	11	9	6	8	5	1	2	3	10
4	9	1	7	6	2	5	10	3	8	11
4	9	7	3	6	10	11	5	2	1	8
4	9	8	2	6	5	10	11	7	1	3
4	10	1	8	6	5	3	7	11	2	9
4	10	5	2	6	8	11	3	7	9	1
4	10	5	8	6	2	11	3	7	9	1
4	10	5	8	6	11	3	2	9	7	1
4	10	7	8	6	11	9	1	5	2	3
4	10	11	2	6	8	5	9	1	3	7
4	10	11	8	6	2	5	9	1	3	7
4	10	11	8	6	5	9	2	3	1	7
4	11	2	8	6	3	10	1	5	9	7
4	11	2	9	6	5	8	10	3	1	7
4	11	5	9	6	2	1	7	10	3	8
4	11	8	2	6	7	3	5	10	1	9
4	11	10	7	6	3	5	1	8	2	9
5	2	1	7	6	4	3	10	9	11	8
5	2	1	8	6	3	10	4	11	7	9
5	2	4	3	6	1	7	11	10	9	8
5	2	8	1	6	4	11	7	3	10	9
5	2	9	11	6	4	1	8	10	3	7
5	3	1	10	6	9	4	2	11	8	7
5	3	2	11	6	1	9	10	8	7	4
5	3	7	1	6	4	9	2	8	10	11

5	3	7	1	6	4	11	10	8	2	9
5	3	11	8	6	1	4	2	9	10	7
5	3	11	9	6	4	1	10	8	2	7
5	4	1	3	6	7	2	11	8	9	10
5	4	1	3	6	8	11	2	7	9	10
5	4	7	3	6	2	11	8	1	9	10
5	4	10	1	6	9	2	3	11	7	8
5	4	11	2	6	3	8	10	1	9	7
5	4	11	2	6	9	10	8	7	1	3
5	8	1	2	6	9	4	10	11	7	3
5	8	1	2	6	11	10	4	9	7	3
5	8	2	1	6	9	4	3	7	11	10
5	8	2	1	6	10	11	4	3	7	9
5	8	3	10	6	1	2	4	7	9	11
5	8	3	11	6	10	1	2	4	9	7
5	8	9	10	6	7	2	4	1	3	11
5	8	10	9	6	1	4	3	11	7	2
5	9	1	4	6	3	7	10	11	2	8
5	9	1	7	6	3	4	2	8	11	10
5	9	1	7	6	4	3	2	8	10	11
5	9	1	10	6	2	7	4	3	11	8
5	9	7	10	6	2	1	4	3	11	8
5	9	8	3	6	4	10	11	7	1	2
5	9	8	11	6	1	3	4	2	7	10
5	9	11	2	6	1	10	8	3	4	7
5	10	1	3	6	8	11	2	4	9	7
5	10	1	9	6	4	3	7	11	2	8
5	10	4	1	6	2	7	11	3	8	9
5	10	4	1	6	9	11	3	8	7	2
5	10	11	8	6	3	4	2	7	1	9
5	11	3	1	6	7	2	4	10	9	8
5	11	3	1	6	9	10	4	2	7	8
5	11	3	7	6	1	2	4	10	9	8
5	11	3	7	6	2	1	4	10	8	9
5	11	3	8	6	1	4	9	2	7	10
5	11	9	2	6	1	10	3	8	7	4
5	11	10	7	6	4	1	3	2	9	8
7	1	2	5	6	8	11	9	10	3	4
7	1	3	10	6	11	2	9	4	5	8
7	1	9	4	6	11	8	3	10	5	2
7	1	9	5	6	10	11	8	2	4	3
7	1	9	5	6	11	10	8	2	3	4
7	1	9	11	6	3	2	8	10	5	4
7	1	9	11	6	5	10	8	2	3	4
7	2	1	4	6	9	8	10	5	11	3
7	2	8	11	6	3	1	9	4	5	10
7	2	8	11	6	10	5	1	9	4	3
7	2	11	3	6	8	9	5	1	10	4
7	2	11	9	6	4	1	10	8	3	5
7	3	1	10	6	11	2	4	9	8	5
7	3	4	1	6	11	9	8	10	5	2
7	3	4	9	6	8	2	1	5	11	10

7	3	5	2	6	10	11	8	9	1	4	8	5	10	11	6	3	7	1	2	4	9
7	3	11	2	6	10	5	8	9	1	4	8	7	1	9	6	10	5	11	2	3	4
7	3	11	5	6	8	9	10	4	2	1	8	7	2	11	6	3	1	5	4	10	9
7	3	11	5	6	9	8	10	4	1	2	8	7	4	10	6	11	3	1	2	5	9
7	3	11	8	6	9	5	2	1	10	4	8	7	10	4	6	3	2	5	1	9	11
7	4	2	3	6	11	8	9	1	5	10	8	7	10	9	6	1	4	2	3	5	11
7	4	3	2	6	5	10	8	11	9	1	8	9	4	10	6	1	2	7	11	5	3
7	4	9	1	6	2	11	10	8	3	5	8	9	5	11	6	10	1	2	3	4	7
7	4	9	2	6	11	10	8	5	3	1	8	9	11	3	6	2	7	1	10	5	4
7	4	10	11	6	2	1	8	9	5	3	8	11	3	5	6	7	10	9	2	4	1
7	4	10	11	6	3	8	9	5	1	2	8	11	3	7	6	4	10	1	2	5	9
7	4	11	10	6	1	2	8	3	5	9	8	11	7	3	6	10	4	9	5	1	2
7	4	11	10	6	3	8	2	1	5	9	8	11	7	9	6	4	1	5	10	3	2
7	8	1	10	6	3	2	4	5	11	9	8	11	10	4	6	7	1	3	2	9	5
7	8	1	10	6	9	4	2	11	3	5	8	11	10	4	6	7	3	1	9	5	2
7	8	2	11	6	3	10	9	1	5	4	8	11	10	5	6	9	1	7	2	4	3
7	8	5	9	6	10	1	4	11	3	2	9	1	3	4	6	8	7	2	11	5	10
7	8	11	9	6	4	1	10	5	3	2	9	1	4	11	6	10	3	7	8	5	2
7	8	11	9	6	5	10	1	4	3	2	9	1	5	2	6	8	7	11	3	4	10
7	9	1	3	6	8	11	2	4	10	5	9	1	5	8	6	2	7	11	3	4	10
7	9	1	4	6	11	8	10	3	2	5	9	1	7	8	6	4	3	2	11	5	10
7	9	5	11	6	8	1	2	4	10	3	9	1	10	5	6	8	2	11	7	3	4
7	9	5	11	6	8	3	10	4	2	1	9	2	4	1	6	10	11	7	3	8	5
7	9	10	1	6	11	3	2	4	5	8	9	2	5	10	6	11	8	7	3	4	1
7	9	11	2	6	3	8	10	1	4	5	9	2	11	4	6	7	1	3	10	8	5
7	10	3	1	6	8	11	4	2	9	5	9	4	1	2	6	7	10	5	3	8	11
7	10	4	11	6	8	1	5	9	2	3	9	4	2	11	6	8	3	10	5	1	7
7	10	8	9	6	11	5	1	2	3	4	9	4	3	7	6	10	11	5	1	2	8
7	10	11	4	6	9	2	8	1	5	3	9	4	3	10	6	7	2	5	1	8	11
7	10	11	5	6	8	9	2	3	1	4	9	4	3	10	6	11	8	2	5	1	7
8	1	2	5	6	9	7	11	4	10	3	9	4	7	2	6	1	10	8	3	5	11
8	1	4	10	6	5	9	7	2	11	3	9	4	7	2	6	11	8	3	5	10	1
8	1	7	3	6	10	11	5	2	9	4	9	5	1	4	6	10	11	7	3	8	2
8	1	10	3	6	7	4	2	9	11	5	9	5	1	10	6	4	11	7	3	8	2
8	1	10	4	6	9	2	11	7	3	5	9	5	2	1	6	4	10	7	11	3	8
8	2	1	4	6	7	3	10	9	11	5	9	5	3	8	6	4	11	10	7	1	2
8	2	1	4	6	10	7	3	11	9	5	9	5	8	7	6	2	3	11	4	1	10
8	2	1	10	6	4	7	3	11	9	5	9	5	11	4	6	8	3	10	7	1	2
8	2	5	4	6	1	3	11	7	10	9	9	7	3	1	6	11	10	4	2	5	8
8	2	7	4	6	1	9	10	3	5	11	9	7	4	11	6	10	3	1	2	5	8
8	2	7	4	6	10	1	9	5	3	11	9	7	4	11	6	10	3	8	5	2	1
8	2	7	10	6	4	1	9	5	3	11	9	7	10	11	6	1	3	4	2	5	8
8	2	11	4	6	7	9	5	1	10	3	9	7	11	2	6	8	3	10	5	1	4
8	3	4	10	6	7	2	1	5	11	9	9	7	11	8	6	2	3	10	5	1	4
8	3	5	9	6	2	1	7	10	11	4	9	8	3	2	6	7	4	10	1	5	11
8	3	11	5	6	10	7	2	9	4	1	9	8	3	2	6	11	10	1	5	4	7
8	5	1	3	6	4	7	11	10	9	2	9	8	3	11	6	2	7	1	5	10	4
8	5	1	9	6	10	4	3	11	7	2	9	8	5	10	6	11	2	1	3	4	7
8	5	9	1	6	4	10	7	2	11	3	9	8	10	7	6	4	3	2	1	5	11
8	5	9	11	6	1	10	3	2	4	7	9	8	11	10	6	5	2	4	3	1	7
8	5	10	4	6	1	7	9	2	3	11	9	8	11	10	6	7	4	3	1	2	5
8	5	10	4	6	1	9	7	3	11	2	9	10	1	2	6	7	4	11	3	8	5

```
 9  10   7   8   6  11   5   3   2   4   1        10  11   8   2   6   1   9   5   4   7   3
 9  10   8   5   6   2   7  11   3   4   1        10  11   8   3   6   1   4   9   7   2   5
 9  11   2   7   6   5   3   8  10   1   4        10  11   8   9   6   5   2   1   3   4   7
 9  11   3   5   6   7   2   8  10   1   4        10  11   8   9   6   7   1   3   4   2   5
 9  11   7   4   6  10   3   2   1   5   8        11   2   3   5   6   4   7   8  10   9   1
 9  11   7  10   6   4   3   2   1   5   8        11   2   4   9   6   7   1   5  10   3   8
 9  11   8   7   6   2   3   4   1  10   5        11   2   7   1   6   4   9  10   3   5   8
 9  11   8   7   6   2   3   5  10   1   4        11   2   7   8   6   9  10   4   5   1   3
10   1   2   8   6   5   9  11   3   7   4        11   2   8   7   6   4   5   1   9  10   3
10   1   3   5   6   7   8  11   2   4   9        11   3   1  10   6   2   7   4   9   5   8
10   1   3   7   6   5   8   9   4   2  11        11   3   5   2   6   7  10   8   9   4   1
10   1   3  11   6   8   5   9   4   7   2        11   3   7   1   6   4   9   2   8  10   5
10   1   5   3   6   8  11   7   2   9   4        11   3   7   1   6   9   4   2   8   5  10
10   1   5   9   6   2   8   3   7  11   4        11   3   7   4   6   9   1  10   5   2   8
10   1   8   7   6   5   3  11   2   4   9        11   3   7  10   6   2   1   4   9   5   8
10   1   8   7   6   9  11   5   4   2   3        11   3   8   5   6   7   9   4   2   1  10
10   1   9   5   6   8  11   3   4   7   2        11   3   8   9   6   4  10   5   1   7   2
10   3   1   7   6   8   5   4   9   2  11        11   4   1   9   6   2   5   8   7   3  10
10   3   4   9   6   7   1   5   2   8  11        11   4   5   2   6   3  10   8   1   7   9
10   3   4   9   6  11   8   1   5   2   7        11   4   5   2   6   9   8  10   7   3   1
10   3   8   5   6  11   9   7   4   2   1        11   4   7   9   6   1   2   5   8   3  10
10   3   8  11   6   5   9   7   4   2   1        11   4   7   9   6   8   5   2   1   3  10
10   4   1   2   6   9   5   8   7   3  11        11   4  10   7   6   3   2   9   5   1   8
10   4   1   8   6   3  11   2   7   9   5        11   5   3   2   6   7  10   9   8   1   4
10   4   3   8   6   9  11   5   1   2   7        11   5   9   1   6   2   7   4  10   8   3
10   4   7   2   6   3  11   8   1   9   5        11   5   9   1   6   7   2   4  10   3   8
10   4   7   8   6   9   5   2   1   3  11        11   5   9   7   6   1   2   4  10   3   8
10   4   9   8   6   3   5  11   7   2   1        11   5   9   7   6   3  10   4   2   1   8
10   5   2   9   6   1   4   3   7   8  11        11   5   9   8   6   7   4   3   2   1  10
10   5   4   1   6   3  11   7   2   8   9        11   5  10   1   6   4   7   9   2   3   8
10   5   4   1   6   3  11   9   8   2   7        11   8   2   7   6   3   4   9   1   5  10
10   5   8   2   6   7   3  11   4   1   9        11   8   2   7   6  10   5   4   9   1   3
10   5   8   3   6   1   7   9   4   2  11        11   8   3  10   6   1   2   4   7   9   5
10   5   8   3   6  11   2   7   9   4   1        11   8   7   2   6   3   4  10   5   1   9
10   5   8   9   6   7   4   3   1   2  11        11   8   7   2   6   5  10   4   3   1   9
10   7   2   8   6  11   3   5   9   1   4        11   8   9   5   6  10   7   2   4   3   1
10   7   3  11   6   8   5   9   4   1   2        11   8   9  10   6   7   2   4   1   3   5
10   7   8   1   6   3   5  11   4   2   9        11   8  10   3   6   7   4   9   5   1   2
10   7   8   1   6  11   9   5   2   4   3        11   9   1   7   6   4   3   2   8  10   5
10   7   9   1   6  11   8   3   4   2   5        11   9   1   7   6   4   5  10   8   2   3
10   7   9   5   6   8  11   3   4   1   2        11   9   2   5   6   7   3  10   8   1   4
10   7   9  11   6   1   8   5   2   4   3        11   9   5   3   6   4   7  10   8   2   1
10   7  11   3   6   2   8   9   1   5   4        11   9   5   8   6   7   4   2   3  10   1
10   7  11   9   6   8   5   1   2   3   4        11   9   7  10   6   3   4   2   5   8   1
10   9   4   3   6   1   7  11   2   8   5        11  10   4   7   6   2   1   5   9   8   3
10   9   4   3   6   5   8   7  11   2   1        11  10   4   7   6   3   5   9   8   1   2
10   9   7   1   6   8  11   4   3   2   5        11  10   5   8   6   9   4   2   1   7   3
10   9   8   5   6  11   3   1   4   2   7        11  10   7   3   6   4   9   1   5   2   8
10   9   8  11   6   5   3   1   4   2   7        11  10   7   9   6   8   5   2   4   3   1
10  11   2   3   6   7   4   9   1   8   5
10  11   4   7   6   9   5   1   2   8   3         1   2   5  11   3   6   7  10   8   9   4
10  11   4   7   6   9   5   3   8   2   1         1   2   7   4   3   6   8   9   5  11  10
```

1	2	7	11	8	6	3	5	9	4	10	3	2	9	7	1	6	4	5	10	8	11
1	2	8	5	10	6	11	3	7	4	9	3	2	11	4	5	6	7	9	10	1	8
1	3	5	2	9	6	8	7	10	4	11	3	2	11	5	4	6	7	8	1	9	10
1	3	7	11	10	6	5	8	2	4	9	3	2	11	9	7	6	5	4	10	1	8
1	3	10	7	8	6	9	2	5	4	11	3	4	1	9	5	6	10	11	8	2	7
1	3	10	8	7	6	4	5	11	2	9	3	4	2	8	11	6	10	5	1	9	7
1	3	10	8	11	6	5	9	2	4	7	3	4	7	2	1	6	10	1i	5	9	8
1	4	2	3	11	6	5	8	10	9	7	3	5	9	4	10	6	1	2	7	11	8
1	4	2	7	8	6	11	5	3	10	9	3	7	1	9	5	6	2	8	11	10	4
1	4	2	9	11	6	5	8	10	3	7	3	7	1	10	5	6	8	9	4	2	11
1	4	10	8	9	6	5	3	11	7	2	3	7	11	8	2	6	1	9	5	4	10
1	4	11	3	7	6	2	5	8	10	9	3	8	5	9	1	6	2	7	4	10	11
1	7	3	11	5	6	8	2	9	10	4	3	8	10	4	7	6	2	1	5	9	11
1	7	9	2	3	6	11	8	10	5	4	3	8	11	10	5	6	2	7	1	9	4
1	8	2	5	4	6	3	10	7	9	11	3	10	1	7	8	6	5	4	2	9	11
1	8	7	3	10	6	9	2	5	11	4	3	10	7	8	1	6	11	9	2	5	4
1	8	7	10	3	6	4	5	2	9	11	3	10	7	9	11	6	1	8	2	5	4
1	8	10	9	7	6	3	2	5	11	4	3	10	8	11	2	6	1	9	5	4	7
1	9	5	4	7	6	3	10	8	11	2	3	10	9	11	5	6	8	1	2	4	7
1	9	5	4	10	6	3	7	11	8	2	3	11	5	2	1	6	4	9	8	10	7
1	9	5	8	3	6	11	10	4	7	2	3	11	5	9	1	6	10	4	7	2	8
1	9	5	11	3	6	8	2	7	4	10	4	1	5	9	2	6	8	3	11	7	10
1	9	10	8	11	6	5	3	2	4	7	4	1	5	9	8	6	2	3	11	7	10
1	10	8	3	4	6	7	2	11	5	9	4	1	5	10	11	6	2	8	3	7	9
1	10	9	2	5	6	8	11	3	7	4	4	1	9	5	8	6	2	11	3	7	10
2	1	4	10	5	6	9	8	11	3	7	4	1	9	5	8	6	11	2	3	10	7
2	1	5	9	11	6	3	8	10	4	7	4	3	7	1	2	6	11	10	5	8	9
2	1	7	3	4	6	9	8	5	10	1	4	3	7	11	2	6	8	5	1	9	10
2	3	11	4	5	6	8	7	1	10	9	4	3	7	11	8	6	2	5	1	9	10
2	3	11	7	4	6	10	1	5	9	8	4	3	8	10	1	6	9	5	11	2	7
2	3	11	7	10	6	4	1	5	9	8	4	5	11	2	3	6	10	9	1	8	7
2	5	1	9	4	6	10	3	7	11	8	4	5	11	2	9	6	1	3	10	8	7
2	5	1	9	10	6	4	3	7	11	8	4	7	3	11	8	6	2	5	9	1	10
2	5	8	10	9	6	1	4	11	3	7	4	7	11	3	2	6	8	9	5	1	10
2	5	9	1	10	6	4	7	3	11	8	4	7	11	3	8	6	2	9	5	1	10
2	7	1	9	4	6	3	8	11	10	5	4	9	1	5	2	6	8	11	7	3	10
2	8	1	9	5	6	10	4	7	11	3	4	9	1	5	8	6	2	11	7	3	10
2	8	3	7	9	6	4	1	5	10	11	4	10	3	2	8	6	11	5	9	1	7
2	8	5	1	9	6	10	4	11	3	7	4	10	5	8	2	6	11	3	7	1	9
2	8	5	10	4	6	9	1	7	3	11	4	11	2	3	5	6	7	8	1	10	9
2	8	7	3	11	6	10	4	1	5	9	4	11	2	8	7	6	5	3	1	10	9
2	8	9	1	3	6	4	7	11	10	5	4	11	10	8	5	6	9	2	3	1	7
2	8	11	10	4	6	3	7	1	9	5	5	2	4	3	8	6	11	10	7	1	9
2	9	5	1	4	6	10	7	11	3	8	5	2	9	10	1	6	4	7	3	11	8
2	9	5	1	10	6	4	7	11	3	8	5	3	1	10	9	6	4	11	2	8	7
2	9	5	4	11	6	8	1	7	10	3	5	3	2	4	7	6	1	9	10	8	11
2	11	3	7	10	6	4	1	9	5	8	5	3	2	4	11	6	8	1	7	10	9
2	11	7	3	4	6	10	9	1	5	8	5	3	2	11	4	6	9	10	1	8	7
2	11	7	3	10	6	4	9	1	5	8	5	3	11	7	2	6	1	4	10	8	9
3	1	9	8	2	6	5	10	11	7	4	5	4	2	9	11	6	3	10	1	7	8
3	2	4	7	10	6	5	9	1	8	11	5	4	10	1	8	6	3	2	11	9	7
3	2	5	11	4	6	1	8	10	9	7	5	4	11	2	3	6	8	1	10	9	7

5	4	11	3	2	6	9	10	1	7	8
5	8	2	4	9	6	1	3	11	7	10
5	8	7	3	11	6	10	1	4	2	9
5	8	10	3	7	6	1	4	2	9	11
5	8	10	9	7	6	1	4	2	3	11
5	8	10	11	4	6	7	1	3	2	9
5	9	1	4	3	6	7	2	8	11	10
5	9	1	8	2	6	3	11	7	4	10
5	9	1	8	11	6	3	2	4	7	10
5	9	2	4	7	6	1	3	10	8	11
5	10	1	7	3	6	11	2	4	9	8
5	10	4	1	2	6	7	3	11	8	9
5	11	3	7	1	6	4	10	9	2	8
5	11	9	10	3	6	7	4	2	1	8
7	2	1	4	9	6	8	3	11	5	10
7	2	1	5	8	6	9	11	3	4	10
7	3	11	5	9	6	8	2	1	4	10

8	1	7	10	3	6	2	9	5	4	11
8	1	7	10	9	6	5	3	2	4	11
8	2	1	4	10	6	7	3	11	5	9
8	2	3	10	4	6	7	1	9	5	11
8	3	4	2	5	6	9	1	7	10	11
8	3	11	5	10	6	7	2	1	4	9
8	5	1	2	7	6	10	4	3	11	9
8	5	9	1	4	6	7	10	3	2	11
8	7	2	4	1	6	9	10	3	5	11
8	7	10	3	1	6	11	4	5	2	9
8	7	10	4	11	6	1	3	5	2	9
9	1	5	8	2	6	7	3	11	4	10
9	2	5	11	4	6	1	8	7	3	10
10	1	4	2	9	6	5	8	7	3	11
10	4	1	5	9	6	2	8	7	3	11
10	5	1	9	7	6	3	4	2	8	11
10	5	8	2	1	6	9	4	7	3	11

SYMMETRY I
Two numbers an equal distance from 6 make a total of 12.

SYMMETRIE I
Zwei Nummern, mit einem gleichen Abstand von 6, ergeben zusammen 12.

1	2	7	4	3	6	9	8	5	10	11		3	1	4	5	10	6	2	7	8	11	9
1	2	7	4	9	6	3	8	5	10	11		3	1	10	5	4	6	8	7	2	11	9
1	3	4	7	2	6	10	5	8	9	11		3	2	5	4	11	6	1	8	7	10	9
1	3	10	7	8	6	4	5	2	9	11		3	2	11	4	5	6	7	8	1	10	9
1	4	3	2	5	6	7	10	9	8	11		3	4	1	2	7	6	5	10	11	8	9
1	4	9	2	5	6	7	10	3	8	11		3	4	7	2	1	6	11	10	5	8	9
1	8	7	10	3	6	9	2	5	4	11		3	5	2	1	8	6	4	11	10	7	9
1	8	7	10	9	6	3	2	5	4	11		3	5	8	1	2	6	10	11	4	7	9
1	9	4	7	2	6	10	5	8	3	11		3	7	4	11	10	6	2	1	8	5	9
1	9	10	7	8	6	4	5	2	3	11		3	7	10	11	4	6	8	1	2	5	9
1	10	3	8	5	6	7	4	9	2	11		3	8	5	10	11	6	1	2	7	4	9
1	10	9	8	5	6	7	4	3	2	11		3	8	11	10	5	6	7	2	1	4	9
2	1	4	3	7	6	5	9	8	11	10		3	10	1	8	7	6	5	4	11	2	9
2	1	4	9	7	6	5	3	8	11	10		3	10	7	8	1	6	11	4	5	2	9
2	1	7	3	4	6	8	9	5	11	10		3	11	2	7	8	6	4	5	10	1	9
2	1	7	9	4	6	8	3	5	11	10		3	11	8	7	2	6	10	5	4	1	9
2	1	8	5	3	6	9	7	4	11	10		4	1	3	5	2	6	10	7	9	11	8
2	1	8	5	9	6	3	7	4	11	10		4	1	3	7	10	6	2	5	9	11	8
2	3	5	11	4	6	8	1	7	9	10		4	1	9	5	2	6	10	7	3	11	8
2	3	11	5	4	6	8	7	1	9	10		4	1	9	7	10	6	2	5	3	11	8
2	5	3	1	4	6	8	11	9	7	10		4	3	1	7	2	6	10	5	11	9	8
2	5	3	11	8	6	4	1	9	7	10		4	3	7	1	2	6	10	11	5	9	8
2	5	9	1	4	6	8	11	3	7	10		4	5	2	3	11	6	1	9	10	7	8
2	5	9	11	8	6	4	1	3	7	10		4	5	2	9	11	6	1	3	10	7	8
2	7	1	3	4	6	8	9	11	5	10		4	5	10	1	3	6	9	11	2	7	8
2	7	1	9	4	6	8	3	11	5	10		4	5	10	1	9	6	3	11	2	7	8
2	7	4	3	1	6	11	9	8	5	10		4	5	11	3	2	6	10	9	1	7	8
2	7	4	9	1	6	11	3	8	5	10		4	5	11	9	2	6	10	3	1	7	8
2	7	8	11	3	6	9	1	4	5	10		4	7	3	1	10	6	2	11	9	5	8
2	7	8	11	9	6	3	1	4	5	10		4	7	3	11	2	6	10	1	9	5	8
2	9	5	11	4	6	8	1	7	3	10		4	7	9	1	10	6	2	11	3	5	8
2	9	11	5	4	6	8	7	1	3	10		4	7	9	11	2	6	10	1	3	5	8
2	11	3	5	8	6	4	7	9	1	10		4	9	1	7	2	6	10	5	11	3	8
2	11	3	7	4	6	8	5	9	1	10		4	9	7	1	2	6	10	11	5	3	8
2	11	9	5	8	6	4	7	3	1	10		4	11	2	3	5	6	7	9	10	1	8
2	11	9	7	4	6	8	5	3	1	10		4	11	2	9	5	6	7	3	10	1	8

4	11	5	3	2	6	10	9	7	1	8
4	11	5	9	2	6	10	3	7	1	8
4	11	10	7	3	6	9	5	2	1	8
4	11	10	7	9	6	3	5	2	1	8
5	2	3	4	1	6	11	8	9	10	7
5	2	9	4	1	6	11	8	3	10	7
5	3	2	11	4	6	8	1	10	9	7
5	3	8	11	10	6	2	1	4	9	7

5	4	11	2	3	6	9	10	1	8	7
5	4	11	2	9	6	3	10	1	8	7
5	8	3	10	1	6	11	2	9	4	7
5	8	9	10	1	6	11	2	3	4	7
5	9	2	11	4	6	8	1	10	3	7
5	9	8	11	10	6	2	1	4	3	7
5	10	11	8	3	6	9	4	1	2	7
5	10	11	8	9	6	3	4	1	2	7

SYMMETRY II

A uniform distance separates any two numbers equaling 12, and
6 intervenes between them

SYMMETRIE II

Zwei Nummern, mit einem gleichen Abstand von einander und mit
6 dazwischen, ergeben zusammen 12

1	2	4	7	3	6	11	10	8	5	9	3	11	7	8	2	6	9	1	5	4	10
1	3	10	8	5	6	11	9	2	4	7	3	11	8	10	5	6	9	1	4	2	7
1	4	2	3	5	6	11	8	10	9	7	4	1	9	5	2	6	8	11	3	7	10
1	4	2	9	5	6	11	8	10	3	7	4	3	1	2	5	6	8	9	11	10	7
1	7	2	4	9	6	11	5	10	8	3	4	7	3	11	2	6	8	5	9	1	10
1	7	8	10	3	6	11	5	4	2	9	4	10	5	1	3	6	8	2	7	11	9
1	8	10	7	9	6	11	4	2	5	3	4	10	11	9	5	6	8	2	1	3	7
1	9	4	7	2	6	11	3	8	5	10	4	11	2	9	5	6	8	1	10	3	7
1	9	7	2	8	6	11	3	5	10	4	5	2	1	3	4	6	7	10	11	9	8
1	9	10	7	8	6	11	3	2	5	4	5	3	2	4	1	6	7	9	10	8	11
1	9	10	8	5	6	11	3	2	4	7	5	8	10	3	1	6	7	4	2	9	11
1	10	5	3	8	6	11	2	7	9	4	5	8	10	9	1	6	7	4	2	3	11
2	1	4	3	7	6	10	11	8	9	5	5	9	2	4	1	6	7	3	10	8	11
2	1	5	8	3	6	10	11	7	4	9	5	9	2	11	4	6	7	3	10	1	8
2	5	9	1	4	6	10	7	3	11	8	5	9	11	10	4	6	7	3	1	2	8
2	7	4	9	1	6	10	5	8	3	11	5	10	8	11	3	6	7	2	4	1	9
2	7	11	8	9	6	10	5	1	4	3	5	11	4	2	3	6	7	1	8	10	9
2	8	1	5	9	6	10	4	11	7	3	7	1	2	4	3	6	5	11	10	8	9
2	8	7	11	3	6	10	4	5	1	9	7	3	4	1	2	6	5	9	8	11	10
2	11	3	7	4	6	10	1	9	5	8	7	8	10	1	3	6	5	4	2	11	9
3	1	5	10	4	6	9	11	7	2	8	8	2	1	5	3	6	4	10	11	7	9
3	1	10	8	7	6	9	11	2	4	5	8	2	7	9	1	6	4	10	5	3	11
3	2	4	11	5	6	9	10	8	1	7	8	3	5	10	1	6	4	9	7	2	11
3	2	11	7	8	6	9	10	1	5	4	8	7	10	9	1	6	4	5	2	3	11
3	4	2	1	7	6	9	8	10	11	5	8	7	11	2	3	6	4	5	1	10	9
3	5	1	2	8	6	9	7	11	10	4	9	4	2	7	1	6	3	8	10	5	11
3	7	4	2	1	6	9	5	8	10	11	9	5	1	8	2	6	3	7	11	4	10
3	8	5	1	2	6	9	4	7	11	10	9	7	10	8	1	6	3	5	2	4	11
3	10	7	11	4	6	9	2	5	1	8	9	8	11	7	2	6	3	4	1	5	10
3	10	8	7	1	6	9	2	4	5	11											

APPLICATION OF KEY-FORMS

The numbers in the work-form should now be applied to our modulations; each number represents the distance measured in half-tones from 0, which may be any tone.

We return to our first key-form example:

1 4 3 2 5 6 7 10 9 8 11

and its first work-form:

0 1 5 8 10 3 9 4 2 11 7 6 (0)

We will call 0 "E flat." The order of modulations is now:

0 1 5 8 10 3 9 4 2 11 7 6 (0)
E♭ E A♭ B D♭ G♭ C G F D B♭ A E♭

Second work-form:

0 11 7 4 2 9 3 8 10 1 5 6 (0)
E♭ D B♭ G F C G♭ B D♭ E A♭ A E♭

Third work-form:

(0) 6 5 1 10 8 3 9 2 4 7 11 0
E♭ A A♭ E D♭ B G♭ C F G B♭ D E♭

Fourth work-form:

(0) 6 7 11 2 4 9 3 10 8 5 1 0
E♭ A B♭ D F G C G♭ D♭ B A♭ E E♭

In the following four illustrations the foregoing key-form with its four work-forms is applied. In each illustration all seven diatonic seventh-chords are introduced. The chord 343 appears in its four places II, III, VI in major and IV in minor, 434 as I and IV in major and VI in minor. Each location gives the chord a different resolution, and we may consider these chords with their resolutions seven different chords. These seven plus the other five seventh-chords total

NUTZANWENDUNG DER GRUND-FORMELN

Die Nummern in der Arbeits-Formel sollten nun auf unsere Modulationen angewandt werden; jede Nummer entspricht der Entfernung, halbtonig gemessen, von 0, welche jeder beliebige Ton sein mag.

Wir kommen auf das erste Beispiel unserer Grund-Formel zurück:

1 4 3 2 5 6 7 10 9 8 11

und auf deren erste Arbeits-Formel:

0 1 5 8 10 3 9 4 2 11 7 6 (0)

Wir wollen 0 Es nennen. Die Ordnung der Modulationen ist dann:

0 1 5 8 10 3 9 4 2 11 7 6 (0)
Es E As H Des Ges C G F D B A Es

Zweite Arbeits-Formel:

0 11 7 4 2 9 3 8 10 1 5 6 (0)
Es D B G F C Ges H Des E As A Es

Dritte Arbeits-Formel:

(0) 6 5 1 10 8 3 9 2 4 7 11 0
Es A As E Des H Ges C F G B D Es

Vierte Arbeits-Formel:

(0) 6 7 11 2 4 9 3 10 8 5 1 0
Es A B D F G C Ges Des H As E Es

In den folgenden vier Beispielen ist die obige Grund-Formel mit ihren vier Arbeits-Formeln angewandt worden. In jedes Beispiel sind alle diatonischen Septimenakkorde eingefügt worden. Der Akkord 343 erscheint in seinen vier Stellungen—II, III, VI in Dur und IV in Moll, 434 als I, IV in Dur und VI in Moll. Jede Lage gibt den Akkorden eine andere Auflösung und wir dürfen diese Akkorde mit ihren sieben Auflösungen als sieben verschiedene Akkorde betrachten. Diese Akkorde und die anderen

twelve, a different chord for each modulation. The order of these chords can be changed 479,001,600 times using only one work-form. As we have 7,708 work-forms, the possibility of variations in these four examples is 3,692,144,332,800.

Sequences are eliminated because no two moves are alike.

1	4	3	2	5	6	7	10	9	8	11	
0	1	5	8	10	3	9	4	2	11	7	6	(0)

e VII$_7^0$ A$\flat$ VII$_7^0$ B II$_7$ D$\flat$ III$_7$ G$\flat$ VI$_7$ c IV$_7$ G V$_7$ F I$_7$ D IV$_7$ b$\flat$ VI$_7$ a III$_7$ e$\flat$ I$_7$

e   As   H   Des   Ges   c   G   F   D   b   a   es

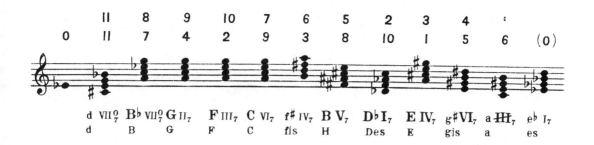

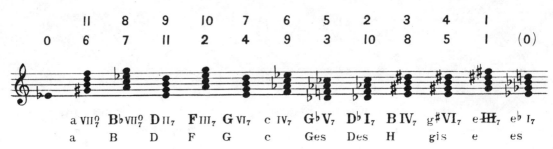

eb a VII$^{o}_{7}$ A$^\flat$ VII$^{o}_{7}$ E II$_{7}$ D$^\flat$ III$_{7}$ B VI$_{7}$ f$^\sharp$ IV$_{7}$ C V$_{7}$ F I$_{7}$ G IV$_{7}$ b$^\flat$ VI$_{7}$ d III$_{7}$ eb I$_{7}$

es a As E Des H fis C F G b d es

434

Nine Chromatic Seventh Chords
and **44** with resolutions 10, 11, 12.

Neun alterierte Septimen
Akkorde und 44 in Auflös -
ungen 10, 11, 12.

SEVENTY NINTH-CHORDS DEVELOPED BY MEANS OF PERMUTATION

A glance at the number-forms of the seventh-chords will show that the possibility of further development has been exhausted.

Diatonic seventh-chords:

333, 334, 343, 433, 434, 443, 334.

Chromatic seventh-chords:

424, 442, 244, 324, 423, 342, 432, 234, 243.

(Each number indicates an interval of a corresponding number of half-tones.)

The ninth-chords are now treated in the same manner. The seventh-chords are the foundations and to each is added a major ninth and a minor ninth—a suggestion from the dominant-ninth-chord in major and in minor. A major ninth is an interval of fourteen half-tones; and a minor ninth, of thirteen. The size of the upper third in the ninth-chord represents the difference between the total of the seventh-chord and the size of the ninth. Thus: if the seventh-chord is 333, the upper third in the ninth-chord will be 5 if the ninth is major and 4 if it is minor.

There is a ninth-chord with an augmented ninth—4344—which is diatonic on the sixth degree in minor.

The symmetric inversion of 3335 is 5333; it is a chord with a major ninth. With a minor ninth the chord would be 5332, and this combination produces twelve new chords.

SIEBZIG NONENAKKORDE ENTWICKELT DURCH PERMUTATION

Ein Blick auf die Zahlenformeln der Septimenakkorde zeigt, dass die Möglichkeiten erschöpft sind.

Diatonische Septimenakkorde:

333, 334, 343, 433, 434, 443, 334.

Chromatische Septimenakkorde:

424, 442, 244, 324, 423, 342, 432, 234, 243.

(Jede Zahl bedeutet ein Intervall einer entsprechenden Zahl von Halbtönen).

Die Nonenakkorde werden genau in derselben Weise behandelt. Die Septimenakkorde bilden die Grundlage, und jedem Akkord wird eine grosse bez. kleine None hinzugefügt—eine Andeutung vom Dominant-Nonenakkord in Dur und Moll. Eine grosse None ist ein Intervall von vierzehn Halbtönen, und eine kleine None ein solches von dreizehn. Der Umfang der oberen Terz im Nonenakkorde bedeutet den Unterschied zwischen der Totalsumme des Septimenakkordes und des Umfanges der None. Deshalb: Wenn der Septimenakkord 333 ist, dann wird die Oberterz in dem Nonenakkord 5, wenn die None gross ist, und 4, wenn sie klein ist.

Es gibt einen Nonenakkord mit übermässiger None—4344—diatonisch auf der 6ten Stufe in Moll.

Die symmetrische Umkehrung von 3335 ist 5333; es ist ein Akkord mit grosser None. Mit kleiner None wird der Akkord 5332 und diese Combination erzeugt zwölf neue Akkorde.

3 10 7 11 4 6 9 2 5 1 8
0 3 1 8 7 11 5 2 4 9 10 6 (0)

	1	4	3	2	5	6	7	10	9	8	11	
0	1	5	8	10	3	9	4	2	11	7	6	(0)

B. ROYT

B. ROYT

B

MODULATION FORMS AND THE SYMMETRIC INVERSION

Symmetry II produces the symmetric inversion:

```
1  2  4  7  3   6  11  10  8  4  9  6  (0)
0  1  3  7  2  5  11  10  8  4  9  6  (0)
```

$1+11, 2+10, 4+8, 7+5, 3+9.$

In the work-form we add 0 to the number below 6 and find a symmetry of 11: $0+11, 1+10, 3+8, 7+4, 2+9, 5+6.$

When the key-form is reversed, we find the work-form 1 below 6, and the symmetry in the work-form will be 1 (or 13):

```
9  5  8  10  11  6  3   7  4  2  1
0  9  2  10   8  7  1  4  11  3  5  6  (0)
```

Changed into letters indicating the keys, the first work-form is:

```
0  1   3   7  2  5  | 11  10  8  4  9   6   (0)
C  D♭  E♭  G  D  F  |  b   b♭  g♯ e  a  f♯  C
```

When we consider the signatures of these keys and add these signatures of a major key and a minor key, we get a total of two sharps: $C+b=2♯$, $D♭+b♭=10♭$ (2♯), $E+g=2♯$, $G+e=2♯$, $D+a=2♯$, $F+f♯=2♯$; and we see that the second half of our work-form is the eleventh symmetric inversion of the first half. (All twelve symmetric inversions have been explained in *Contributions to Theory* by Thorvald Otterström.) In the work-form developed from the reversed key-form, 1 is below 6. The keys are:

```
0  9  2  10   8  7  | 1   4  11  3  5   6   (0)
C  A  D  B♭  A♭  G  | c♯  e   b  d♯ f  f♯  C
```

MODULATIONSFORMELN UND DIE SYMMETRISCHE UMKEHRUNG

Symmetrie II produziert die symmetrische Umkehrung:

```
1  2  4  7  3   6  11  10  8  4  9  6  (0)
0  1  3  7  2  5  11  10  8  4  9  6  (0)
```

$1+11, 2+10, 4+8, 7+5, 3+9.$

In der Arbeits-Formel addieren wir 0 zu der Ziffer unter 6 und finden eine Symmetrie von 11: $0+11, 1+10, 3+8, 7+4, 2+9, 5+6.$

Wenn die Grund-Formel umgekehrt wird, finden wir in der Arbeits-Formel 1 unter 6 und die Symmetrie in der Arbeits-Formel wird 1 (oder 13) sein:

```
9  5  8  10  11  6  3·  7  4  2  1
0  9  2  10   8  7  1  4  11  3  5  6  (0)
```

In Buchstaben ausgedrückt—die Tonarten anzeigend—ist die erste Arbeits-Formel:

```
0   1    3   7  2  5  | 11  10  8  4  9   6   (0)
C  Des  Es  G  D  F  |  h   b  gis e  a  fis  C
```

Betrachten wir die Vorzeichnungen dieser Tonarten und zählen wir diese Vorzeichnungen von einer Dur Tonart und einer Moll Tonart zusammen, so erhalten wir die Summe von zwei Kreuzen: $C+h=2♯$; $Des+b=10♭$ (2♯); $E+g=2♯$; $G+e=2♯$; $D+a=2♯$; $F+fis=2♯$ und wir sehen dass die zweite Hälfte unserer Arbeits-Formel die elfte symmetrische Umkehrung der ersten Hälfte ist. (Alle zwölf symmetrischen Umkehrungen sind in *Beiträge zur Theorie* von Thorvald Otterström erläutert worden.) In der aus der umgekehrten Grund-Formel entwickelten Arbeits-Formel ist 1 unter 6. Die Tonarten sind:

```
0  9  2  10   8  7  | 1    4  11  3  5   6   (0)
C  A  D  B  As  G  | cis  e   h  dis f  fis  C
```

The total is four sharps: C+c♯, A+e,
D+b, B♭+d♯, A♭+f, G+f♯, which is
the first symmetric inversion.

With other keys as starting-points we
have this result:

C+b total 2♯. Sym. inv. No. 11
D♭+c total 4♯. Sym. inv. No. 1
D+c♯ total 6♯. Sym. inv. No. 3
E♭+d total 4♭. Sym. inv. No. 5
E+d♯ total 2♭. Sym. inv. No. 7
F+e total 0. Sym. inv. No. 9
F♯+f total 2♯. Sym. inv. No. 11
G+f♯ total 4♯. Sym. inv. No. 1
A♭+g total 6♭. Sym. inv. No. 3
A+g♯ total 4♭. Sym. inv. No. 5
B♭+a total 2♭. Sym. inv. No. 7
B+b♭ total 0. Sym. inv. No. 9

With 1 below 6 the result is:

C+c♯ total 4♯. Sym. inv. No. 1
D♭+d total 6♭. Sym. inv. No. 3
D+d♯ total 4♭. Sym. inv. No. 5
E♭+e total 2♭. Sym. inv. No. 7
E+f total 0. Sym. inv. No. 9
F+f♯ total 2♯. Sym. inv. No. 11

Proceeding farther up the scale, the
totals and the inversions will be repeated.

The same six inversions will appear
when key-form and work-form show:

6 6 6 6
9 3 7 5

Die Totalsumme ist vier Kreuze: C+
cis, A+e, D+h, B+dis, As+f, G+fis,
was .der ersten symmetrischen Umkehrung entspricht.

Von anderen Tonarten als Ausgangs-
punkt erreichen wir dieses Resultat:

C+h total 2♯. Symm. Umk. Nr. 11
Des+c total 4♯. Symm. Umk. Nr. 1
D+cis total 6♯. Symm. Umk. Nr. 3
Es+d total 4♭. Symm. Umk. Nr. 5
E+dis total 2♭. Symm. Umk. Nr. 7
F+e total 0. Symm. Umk. Nr. 9
Fis+f total 2♯. Symm. Umk. Nr. 11
G+fis total 4♯. Symm. Umk. Nr. 1
As+g total 6♭. Symm. Umk. Nr. 3
A+gis total 4♭. Symm. Umk. Nr. 5
B+a total 2♭. Symm. Umk. Nr. 7
H+b total 0. Symm. Umk. Nr. 9

Mit 1 unter 6 ist das Resultat:

C+cis total 4♯. Symm. Umk. Nr. 1
Des+d total 6♭. Symm. Umk. Nr. 3
D+dis total 4♭. Symm. Umk. Nr. 5
Es+e total 2♭. Symm. Umk. Nr. 7
E+f total 0. Symm. Umk. Nr. 9
F+fis total 2♯. Symm. Umk. Nr. 11

Die Tonleiter weiter aufwärts verfol-
gend wiederholen sich die Totalsummen
und Umkehrungen.

Dieselben sechs Umkehrungen ergeben
sich wenn Grund-Formel und Arbeits-
Formel diese Ziffern zeigen:

6 6 6 6
9 3 7 5

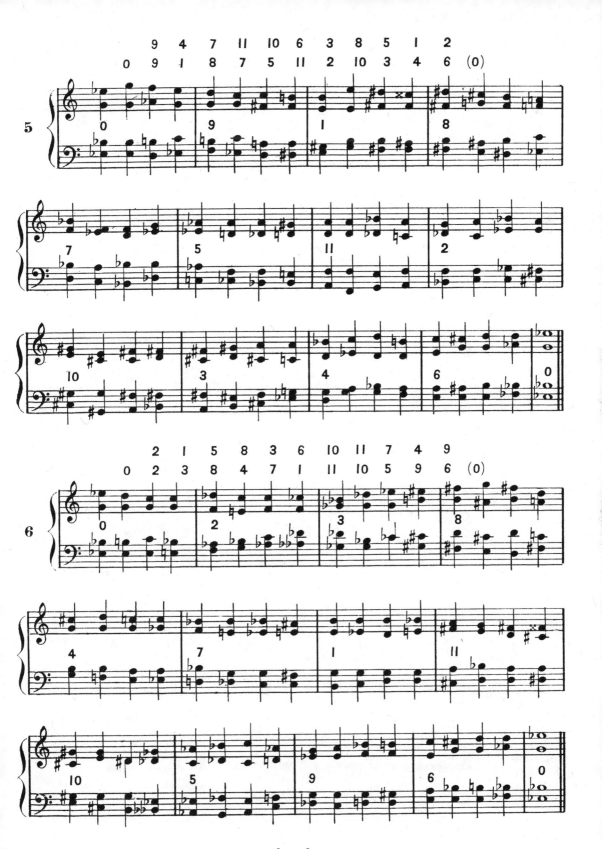

THE DIMINISHED SEVENTH-CHORD

Although it might seem anachronistic to make remarks on this chord in our day (1935) of musical culture, nevertheless there may be some features which have escaped observers so far.

A German-American teacher of theory, whose verbosity by far transcended his erudition, called the dominant seventh-chord "the sacred chord." The man should have been a poet; he would undoubtedly have written poetry in the style of the "Lost Chord," where adjectives supposedly enhance the value of the nouns. In scientific works adjectives are superfluous.

But if there is a chord entitled to a mark of distinction, it is the diminished seventh-chord—this humble, much used, and still more abused chord. It is the only seventh-chord in which all the thirds are equal.

1. By means of orthographical variation it reproduces itself in eleven keys.

2. Its connection with the tonic major triad unites two epochs of music separated by centuries: the ecclesiastical mode and the symmetric inversion.

3. By means of half-tone progressions it becomes the parent of all diatonic and chromatic seventh-chords, and through these it connects with all major and minor modes.

1

f: VII⁰₇   g: VII⁰₇   b: VII⁰₇   d: VII⁰₇

DER VERMINDERTE SEPTIMENAKKORD

Obwohl es in unserem Zeitalter (1935) überflüssig erscheinen mag, Bemerkungen über diesen Akkord zu machen, so mögen·doch einige Tatsachen erwähnt werden, welche—wie es scheint—der Beobachtung entgangen sind.

Ein deutsch-amerikanischer Theorielehrer—dessen äusserlicher Wortreichtum sein Wissen weit übertraf—bezeichnete den Dominant-Septimenakkord als "den heiligen Akkord." Dieser Mann hätte Dichter sein sollen; zweifellos würde er Poesie im Style des "Lost Chord" geschrieben haben. "Wo Begriffe fehlen, stellt zur rechten Zeit ein Wort sich ein" (Goethe, *Faust*). In wissenschaftlichen Werken sind Adjective überflüssig.

Nun gibt es einen Akkord, der diese Auszeichnung wirklich verdienen könnte—der verminderte Septimenakkord, dieser bescheidene, vielgebrauchte und vielgeschmähte Akkord. Es ist der einzige Septimenakkord, in welchem alle Terzen gleich sind.

1. Orthographische Variation ermöglicht seine Darstellung in elf Tonarten.

2. Seine Verbindung mit dem tonischen Dreiklang (Dur) verbindet zwei Musikepochen (durch Jahrhunderte von einander getrennt): den kirchlichen Modus und die symmetrische Umkehrung.

3. Durch Anwendung der Halbtonfortschreitung wird er der Erzeuger aller diatonischen und chromatischen Septimenakkorde, und durch diese verbindet er sämtliche Dur- und Moll-Skalen.

1

f: VII⁰₇   gis: VII⁰₇   h: VII⁰₇   d: VII⁰₇

The tonics of these keys form another diminished seventh-chord.

Die ersten Stufen dieser Skalen formen einen andern verminderten Septimenakkord.

Between the root position and the first inversion of the diminished seventh-chord are two other diminished seventh-chords. The tonics of *a* and *b* furnish these two chords. The tonics of *c* equal acoustically the first inversion of the first chord in *a*.

Zwischen der Grundstellung und der ersten Umkehrung des verminderten Septimenakkordes liegen zwei andere verminderte Septimenakkorde. Die ersten Stufen von *a* und *b* formen diese Akkorde. Die ersten Stufen von *c* sind akustisch die erste Umkehrung des ersten Akkordes von *a*.

2

The diminished seventh-chord and its tonic major resolution are to be located in one key:

Der verminderte Septimenakkord und seine tonische Dur-Auflösung sollen in einer Tonart auftreten:

It is not C major because C major has no A flat. It is not C minor because C minor has no E.

Es ist nicht C Dur, weil C Dur kein As hat. Es ist nicht C Moll, weil C Moll kein E hat.

We must go back to the old aeolian scale and its plagal scale, the hypo-aeolian scale, and their symmetric inversions.

Wir müssen zurückgehen zur alten aeolischen Tonart und deren Plagaltonart, der hypo-aeolischen und deren symmetrischen Umkehrungen.

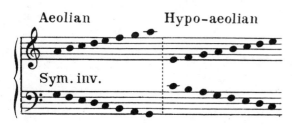

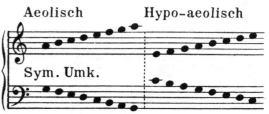

We adopted this scale for our harmonic minor and raised the seventh degree. We may call it the "neo-aeolian scale."

Man machte diese Skala zur harmonischen Molltonleiter und erhöhte die siebente Stufe. Man mag sie die neo-aeolische Skala nennen.

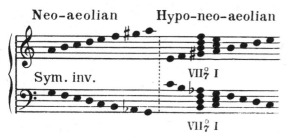

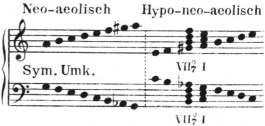

It must be remembered that the tonic in the plagal scale is the same as in the authentic. In the symmetric inversion the degrees are descending.

Man muss sich vergegenwärtigen, dass die Tonika in der plagalen Skala dieselbe ist wie die Tonika in der authentischen Tonart. In der symmetrischen Umkehrung sind die Stufen absteigend.

The cadence

Die Kadenz

is therefore vii°₇ I in the symmetric inversion of the hypo-neo-aeolian scale.

ist daher vii°₇ I in der symmetrischen Umkehrung von der hypo-neo-aeolischen Skala.

3

It is possible to make thirty-three half-tone moves from the diminished seventh-chord without moving all four notes. In the illustrations the static notes are expressed in whole notes, while the moving notes are quarters.

Es ist möglich dreiunddreissig Halbtonfortschreitungen von dem verminderten Septimenakkord herzustellen, ohne alle vier Töne fortschreiten zu lassen. In den Beispielen sind die liegenbleibenden Töne durch Ganznoten ausgedrückt (im Gegensatz zu den fortschreitenden Viertelnoten).

ON SCALES

We have terms like: "pentatonic (five-tone) scale," "hexatonic (six-tone) scale," "heptatonic (seven-tone) scale," and "chromatic scale." The definition of the word "scale" in music is: tones progressing by adjacent degrees. We impose this definition on ourselves by the use of the double sharp; we use F$\times$ in G$\sharp$ minor because F$\times$ is the adjacent degree to the preceding E while G is a leap—a break in the scale. G to G$\sharp$ is not an adjacent degree; it is the same degree raised. With this definition as a basis the scale can have only seven tones, because our staff of five lines has only seven "degrees" before we reach the octave, and it can have neither more nor less than seven.

In the pentatonic scale two degrees are missing; and in the hexatonic, one; while five degrees are repeated in the chromatic scale. In the strict sense of the term these progressions do not deserve the name "scale"; but unfortunately we have no other identification which we can apply to them.

In other respects our terminology needs revision before the theory of music can be considered an exact science like mathematics.

We use the terms "sharp" (meaning single sharp) and "double sharp." The double sharp is a single cross and the single sharp has four crosses!

We divide the quarter note into two eighths, four sixteenths, eight thirty-seconds, etc. with no division between adjacent groups. We divide the quarter note into three parts and still we call them "eighths," while in reality they are twelfths. Each of these "eighths" has two sixteenths but can also have three

ÜBER SKALEN

Wir haben Bezeichnungen wie: "Pentatonische (fünftonig)," "Hexatonische (sechstonig)," "Heptatonische (siebentonig) Skala," und "Chromatische Skala." Die Definition des Wortes "Skala" ist: "Töne in stufenweiser Fortschreitung." Das doppelte Erhöhungszeichen ($\times$) ist der Grund dieser Definition. Wir gebrauchen fis-is in Gis Moll, weil fis-is die benachbarte Stufe zum vorangegangenen E ist, während G einen Sprung bedeutet—eine Kluft in der Skala. G-gis ist keine Stufenfortschreitung, sondern dieselbe Stufe chromatisch erhöht. Mit dieser Definition als Basis kann die Skala nur sieben Töne haben, weil unser Fünfliniensystem nur sieben Stufen hat bis wir die Octave erreichen, und es kann weder mehr noch weniger als sieben Töne haben.

In der Pentatonischen Skala fehlen zwei Stufen und in der Hexatonischen eine, während in der chromatischen Skala fünf Stufen wiederholt werden. Genau genommen verdienen diese Fortschreitungen nicht den Namen Skala; leider haben wir bis jetzt keine passende wissenschaftliche Bezeichnung.

In anderer Beziehung erfordert unsere Terminologie Revision, ehe die Musik als eine exacte Wissenschaft bezeichnet werden kann, wie z.B. die Mathematik.

Wir sagen "Kreuz" und "Doppelkreuz." Das doppelte Kreuz ist ein einfaches Kreuz, und das einfache Kreuz hat vier Kreuze!

Wir teilen die Viertel-Note in zwei Achtel, vier Sechzehntel, acht Zweiunddreissigstel, etc. ohne Abteilungen zwischen benachbarten Gruppen. Wir teilen die Viertel-Note in drei Teile und nennen sie trotzdem Achtel-Noten, obwohl es Zwölftel-Noten sind. Jedes von diesen Achteln hat zwei Sechzehntel, kann aber auch drei haben (Triolen) und die

(triplets), and the nine notes remain sixteenths; hence the curious notation in
Siegfried's Funeral March from Wagner's *Götterdämmerung:*

which proves a sixteenth to be worth less
than a thirty-second. Of course, the
"sixteenths" are thirty-sixths, but we
have no term for this division.

We say that a seven-tone scale has
seven degrees; but the first tone of the
scale cannot be a degree because it is the
starting-point and degrees indicate dis-
tances from the starting-point. We
need only to look at a thermometer, a
yardstick, and our globe to realize that
all degrees are counted from zero. A
seven-tone scale can therefore have only
six degrees.

Whether the use of the Roman numerals in designating the degrees is responsi-
ble for this error is a matter of conjec-
ture. The Roman numerals have no sign
for zero; we are indebted to the Arabs
for this symbol.

The Lydian mode was the first scale
we lost in the early Middle Ages. All
church music was vocal, and the singers
could not intonate the augmented fourth
(*tritone*, "*diabolus in musica*"), nor could
they sing three whole tones in succession.
"As they made the sign of the cross before the real devil, so they made the sign
of the flat (♭) before this music devil"
(Angul Hammerich, *Studies in Icelandic
Music* [Copenhagen, 1900]).

There is something elemental about
the Lydian mode which none of the other
modes has except the rejected Locrian
mode, which is perhaps still more elemental. The Lydian mode is quite frequent
in Norwegian folk songs and dances; and
in Iceland it is, or was, used so much

neun Noten bleiben Sechzehntel; daher
die sonderbare Schreibweise in dem
Trauermarsch aus Wagner's Götterdämmerung:

ein Beweis, dass ein Sechzehntel einen
geringeren Zeitwert hat als ein Zweiunddreissigstel. Natürlich sind die Sechzehntel Sechsunddreissigstel; aber wir
haben dafür keine Bezeichnung.

Wir sagen, dass eine siebentonige
Skala sieben Stufen hat; doch der erste
Ton einer Skala ist keine Stufe, weil er
der Anfangston ist, und die Stufen be-
deuten Fortschreitungen vom Anfangstone aus. Wir brauchen bloss ein Thermometer, eine Elle, einen Globus zu
betrachten, um uns zu überzeugen, dass
alle Stufen von Null (0) aus berechnet
sind. Eine siebentonige Skala kann deshalb nur sechs Stufen haben.

Ob der Gebrauch der römischen Zahlen
in Analyse-Berechnung für den Irrtum
verantwortlich ist, mag dahingestellt
bleiben. Die römischen Ziffern haben
keine Bezeichnung für 0; dieses Symbol
verdanken wir den Arabern.

Die lydische Tonart war die erste
Skala, welche im frühen Mittelalter
verloren ging. Alle Kirchenmusik war
vokalen Characters, und die Sänger
konnten eine übermässige Quarte (Tritonus—mit der damaligen Bezeichnung
"diabolus in musica") nicht intonieren;
auch eine Folge von drei Ganztönen war
unzulässig. "Wie man das Kreuzzeichen
(♯) vor dem wirklichen Teufel machte,
so machte man das B-zeichen (♭) vor
diesem Musikteufel" (Angul Hammerich, *Studien in Isländischer Musik* [Ko-
penhagen, 1900]).

Es ist etwas Ursprüngliches in der
lydischen Tonart, welches nicht in andern Tonarten zu finden ist, mit Ausnahme der verfehmten lokrischen Tonart, welche vielleicht noch ursprüng-
licher ist. Die lydische Tonart findet
man häufig in norwegischen Volkslie-
dern und Tänzen, und in Island war sie

that Hammerich suggests calling it the "Icelandic mode."

It is encouraging to note that vocalists have made great intellectual progress since the disappearance of the Lydian mode with its three successive whole tones, for now they have four successive whole tones in the so-called "melodic minor scale"—which was also a vocal accommodation because they could not intonate the augmented second in the harmonic minor scale.

so im Gebrauch, dass Hammerich vorschlug sie die isländische Tonart zu nennen.

Es ist erfreulich zu beobachten, dass Sänger bedeutende intellectuelle Fortschritte seit dem Verschwinden der lydischen Tonart mit ihren drei Ganztonfortschreitungen gemacht haben; denn jetzt haben sie vier Ganztonfortschreitungen in der sogenannten melodischen Moll-Skala — auch eine Concession an die Sänger, welche unfähig waren, die übermässige Sekunde in der harmonischen Moll-Skala zu intonieren.

PERMUTATION APPLIED TO SCALES

Our major and minor scales were evolved from the medieval system of ecclesiastical modes. We have only two scales—major and minor; if the Hungarian scale is accepted, we have two minor scales, making a total of three scales as a basis for most of the music written since Bach.

The previous system had six scales: Ionian, Dorian, Phrygian, Lydian, Mixolydian, and Aeolian; a seventh mode— the Locrian—was developed and rejected. Nothwithstanding, a melody in the Locrian mode is quoted by Hammerich.

For our investigation we will convert these seven modes into numbers:

2212221 Ionian
2122212 Dorian
1222122 Phrygian
2221221 Lydian
2212212 Mixolydian
2122122 Aeolian
1221222 Locrian

In the Middle Ages the scales were often transposed; hence the name G-Dorian, C-Mixolydian, etc. We will do the same: transpose all the scales except the Ionian and have them all start from the same tone:

2212221
2122212
1222122
2221221
2212212
2122122
1221222

It is now becoming evident that the difference in the seven scales is caused by permutation because all the factors in all of them are the same: five 2's and

PERMUTATION AUF SKALEN ANGEWANDT

Unsere Dur und Moll Skalen wurden aus den mittelalterlichen Kirchentonarten entwickelt. Wir haben nur zwei Skalen—Dur und Moll; falls die ungarische Tonleiter mitgerechnet wird, haben wir zwei Moll Skalen, insgesamt drei Skalen als Grundlage der meisten seit Bach geschaffenen Musik.

Das frühere System hatte sechs Skalen (jonisch, dorisch, phrygisch, lydisch, mixolydisch und aeolisch). Eine siebente Tonart—die lokrische—wurde entwickelt und verworfen. Dessenungeachtet existiert eine Melodie in dieser Tonart, angeführt von Hammerich.

Für unsere Untersuchung werden wir die Skalen in Ziffern angeben:

2212221 jonisch
2122212 dorisch
1222122 phrygisch
2221221 lydisch
2212212 mixolydisch
2122122 aeolisch
1221222 lokrisch

Im Mittelalter wurden die Skalen öfter transponiert; daher die Bezeichnungen G-dorisch, C-mixolydisch, etc. Wir folgen diesem Vorgange, indem wir alle Skalen, mit Ausnahme der jonischen, transponieren, und zwar vom selben Tone aus:

2212221
2122212
1222122
2221221
2212212
2122122
1221222

Es wird jetzt klar, dass der Unterschied der sieben Skalen durch Permutation hervorgebracht wird, weil die Bestandteile dieselben sind: fünf mal 2 und zwei mal 1. Für unsere nächste Unter-

two 1's. For our next experiment we will extend the scales through three octaves:

```
2212221221222122122221
2122212212221221222122
1222122122211221222122
2221221222212212221221
2212221222122122222212
2122122221221222122122
1221222212212212212212212
```

In this arrangement the scales may be read horizontally from left to right and vertically from above and down. When this reading is reversed, we have the symmetric inversion according to which the

Ionian 2212221 becomes Phrygian 1222122
Dorian 2122212 remains Dorian 2122212
Phrygian 1222122 becomes Ionian 2212221
Lydian 2221221 becomes Locrian 1221222
Mixolydian 2212212 becomes Aeolian 2122122
Aeolian 2122122 becomes Mixolydian 2212212
Locrian 1221222 becomes Lydian 2221221

If we read seven numbers vertically up and then continue horizontally, the horizontal reading is the symmetric inversion of the vertical. We may also read the upper line (seven numbers) backward and then continue down and get the symmetric inversion of the horizontal line. Or we may begin in the upper line from the left, drop anywhere into the second line, and go either horizontally or vertically—it does not interfere with the scale. Or we may select one corner of the diagram (the lower left), start in the upper right corner, and move in the outside numbers only and in groups of seven. It will take us twelve times around before we arrive at the starting-point as the first number of a scale; and in the twelve scales we will find our seven ecclesiastical modes.

These curiosities belong in the realm of amusement and are of no importance for our investigation. Far more important is the diagonal reading. From right

suchung erweitern wir die Skalen auf drei Octaven:

```
2212221221222122122221
2122212212221221222122
1222122122211221222122
2221221222212212221221
2212221222122122222212
2122122221221222122122
1221222212212212221222
```

In dieser Anordnung können die Skalen horizontal von links nach rechts und vertikal von oben nach unten gelesen werden. Wenn diese Lesart umgekehrt wird, haben wir die symmetrische Umkehrung, nach welcher aus

jonisch 2212221 phrygisch 1222122 wird
dorisch 2122212 dorisch 2122212 bleibt
phrygisch 1222122 jonisch 2212221 wird
lydisch 2221221 lokrisch 1221222 wird
mixolydisch 2212212 aeolisch 2122122 wird
aeolisch 2122122 mixolydisch 2212212 wird
lokrisch 1221222 lydisch 2221221 wird

Lesen wird die sieben Nummern vertikal aufwärts und fahren fort horizontal, dann ist die horizontale Lesart die symmetrische Umkehrung der vertikalen. Wir könnten auch die obere Zeile (sieben Nummern) rückwärts lesen und dann abwärts fortfahren, und wir behalten die symmetrische Umkehrung der horizontalen Zeile. Oder wir könnten mit der oberen Zeile links anfangen und springen irgendwo in die zweite Zeile, bewegen uns entweder horizontal oder vertikal—die Skala bleibt unverändert. Oder wir könnten eine Ecke des Vierecks wählen (die untere linke), fangen in der rechten oberen Ecke an und bewegen uns ausschliesslich in den äusseren Ziffern und in Gruppen von sieben. Diesen Weg müssen wir zwölf mal zurücklegen, bis wir den Anfangspunkt als erste Nummer von einer Skala erreichen. Und in den zwölf Skalen finden wir unsere sieben Kirchentonarten.

Diese Eigentümlichkeiten gehören in das Reich der Spielereien und sind belanglos für unsere Untersuchungen. Weit wichtiger ist die diagonale Lesart. Von rechts nach links abwärts (oder von

to left down (or left to right up) we find the number form for the hexatonic (whole-tone) scale (which was "invented" or "discovered" in France many centuries later)—222222, and we also find the first seven tones of the chromatic scale. The rest will appear when we start the scales over again under the Locrian—beginning with the Ionian; the last will be the Mixolydian and the diagonal line will show twelve 1's—the entire chromatic scale.

If we read diagonally from left to right down (or vice versa), we will get seven new scales in which the factors are identical with those of the original seven scales —five 2's and two 1's; but the arrangement will be different. Here the permutation has taken place automatically.

We now have fourteen scales; and when we want to exhaust the possibilities and find out how many there are, the most positive procedure is by way of mathematics.

There are seven factors in the scale which equal $7 \times 6 \times 5 \times 4 \times 3 \times 2 \times 1$; since five 2's and two 1's are found in these number-forms, we must divide with $(5 \times 4 \times 3 \times 2 \times 1) \times (2 \times 1)$.

$$\frac{7 \times 6 \times 5 \times 4 \times 3 \times 2 \times 1}{(5 \times 4 \times 3 \times 2 \times 1) \times (2 \times 1)} = \frac{5040}{240} = 21$$

Now we know that our major scale produces twenty-one scales.

The harmonic minor in numbers is 2122131. The permutation is

$$\frac{5040}{36} = 140 .$$

The Hungarian scale is 2131131 and in permutation

$$\frac{5040}{48} = 105 .$$

links nach rechts aufwärts) finden wir die Zahlenformel der hexatonischen (ganztonigen) Skala, (welche Jahrhunderte später in Frankreich entweder "erfunden" oder "entdeckt" wurde)— 222222. Auch finden wir die ersten sieben Töne der chromatischen Tonleiter. Die noch übrigen Töne werden wieder erscheinen, wenn wir die Skala noch einmal wiederholen unter der lokrischen—mit der jonischen anfangen; die letzte ist die mixolydische und die diagonale Linie wird zwölf mal 1 zeigen —die ganze chromatische Skala.

Wenn wir diagonal von links nach rechts abwärts lesen (oder umgekehrt), erhalten wir sieben neue Skalen, in welchen die Bestandteile dieselben sind wie in den sieben Ur-Skalen—fünf mal 2 und zwei mal 1, aber das Arrangement ist anders. Die Permutation fand hier automatisch statt.

Wir haben jetzt vierzehn Skalen, und wenn wir die Möglichkeiten erschöpfen und ihre Totalsumme ausfindig machen wollen, so ist der sicherste Weg der mathematische.

Die Skala hat sieben Bestandteile; die Permutation ist $7 \times 6 \times 5 \times 4 \times 3 \times 2 \times 1$; da wir in diesen Formeln die Zahl 2 fünf mal und die Zahl 1 zwei mal finden, müssen wir mit $(5 \times 4 \times 3 \times 2 \times 1) \times (2 \times 1)$ dividieren.

$$\frac{7 \times 6 \times 5 \times 4 \times 3 \times 2 \times 1}{(5 \times 4 \times 3 \times 2 \times 1) \times (2 \times 1)} = \frac{5040}{240} = 21$$

Jetzt wissen wir dass unsere Dur Skala einundzwanzig Skalen hervorbringt.

Die harmonische Moll Skala in Ziffern ausgedrückt ist: 2122131. Die Permutation ist

$$\frac{5040}{36} = 140 .$$

Die ungarische Skala ist 2131131 und in Permutation

$$\frac{5040}{48} = 105 .$$

Total result from these three scales:

$$21 + 140 + 105 = 266 .$$

These scales will now be presented all starting from the same tone C. This is sometimes changed into B sharp or D double flat in order to avoid the necessity of using triple sharps and triple flats.

Totalsumme von diesen Skalen:

$$21 + 140 + 105 = 266 .$$

Diese Skalen erscheinen jetzt alle vom Anfangstone C aus, welcher mitunter als His oder Des-es notiert wird, um dreifache Kreuze und Been zu vermeiden.

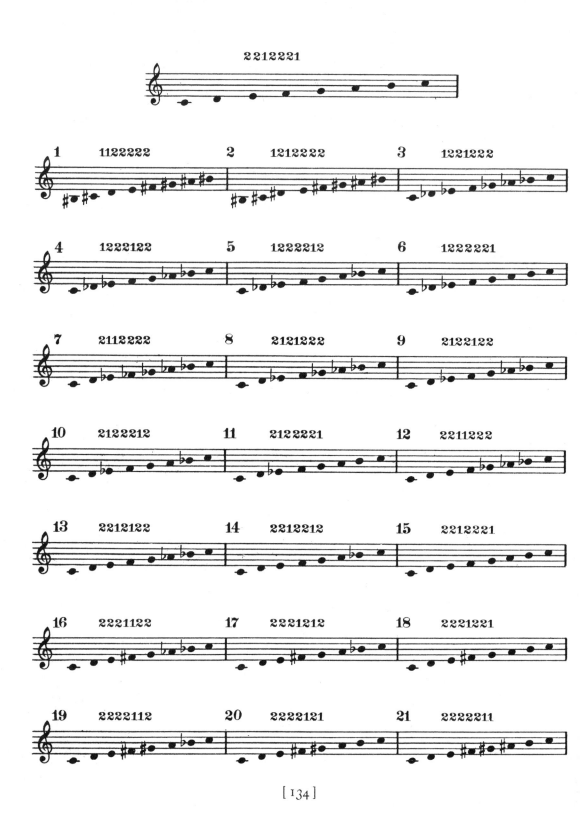

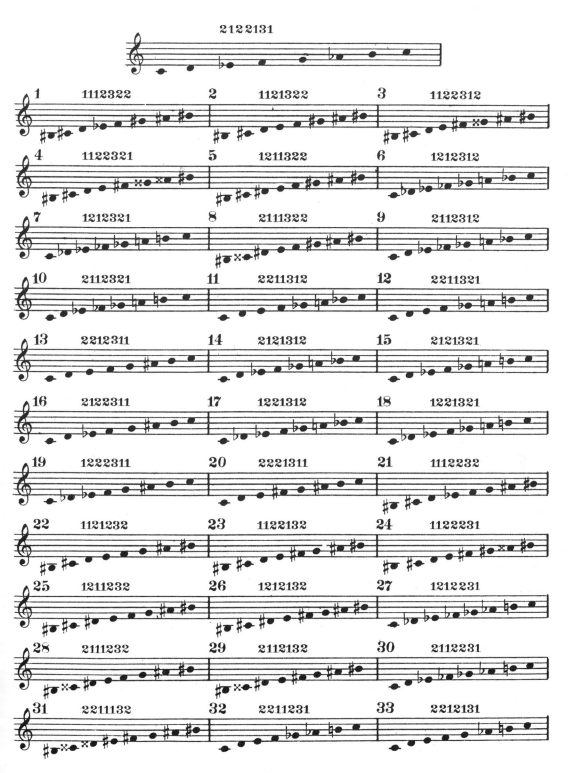

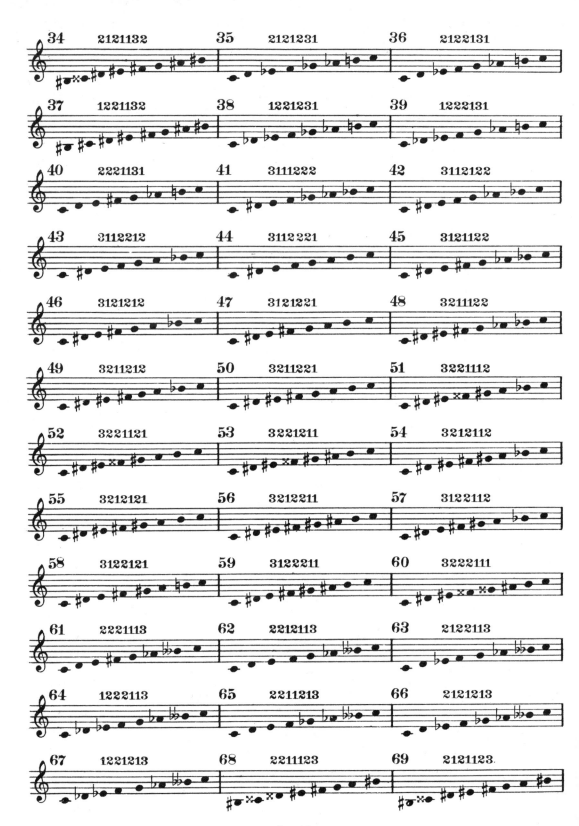

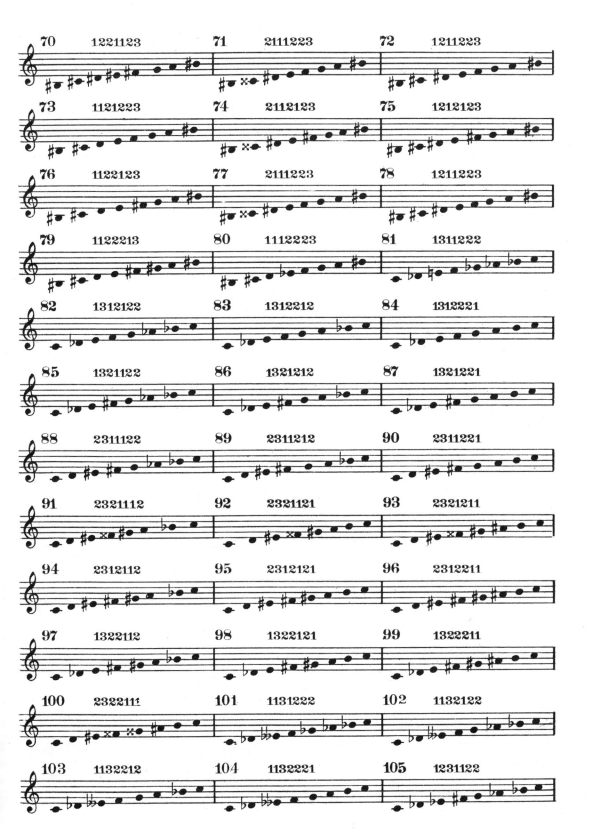

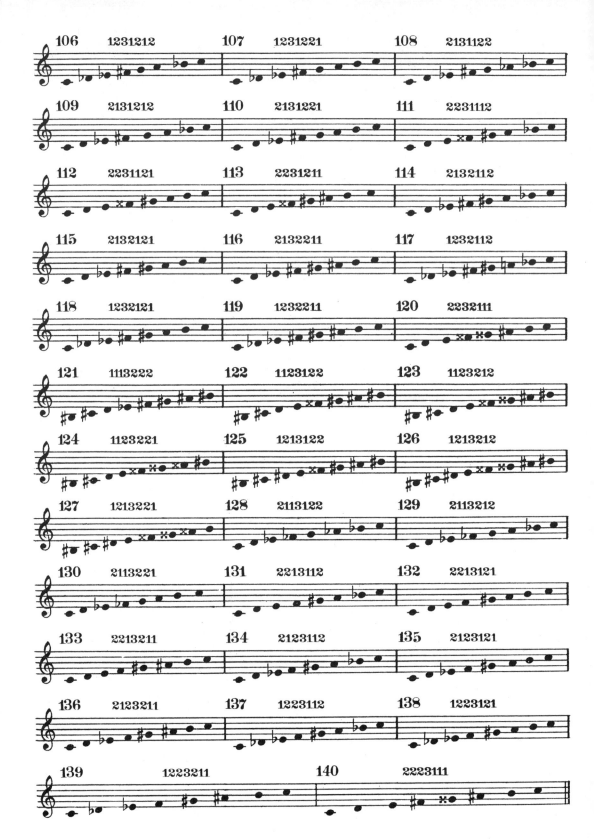

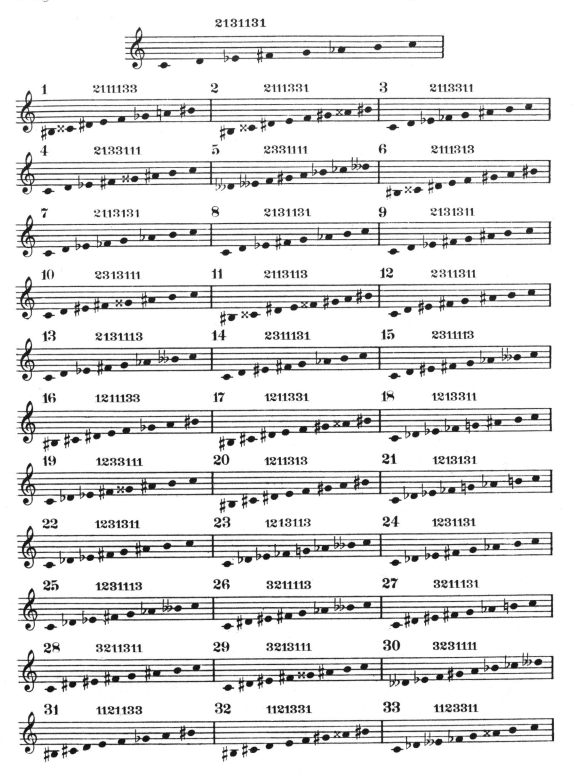

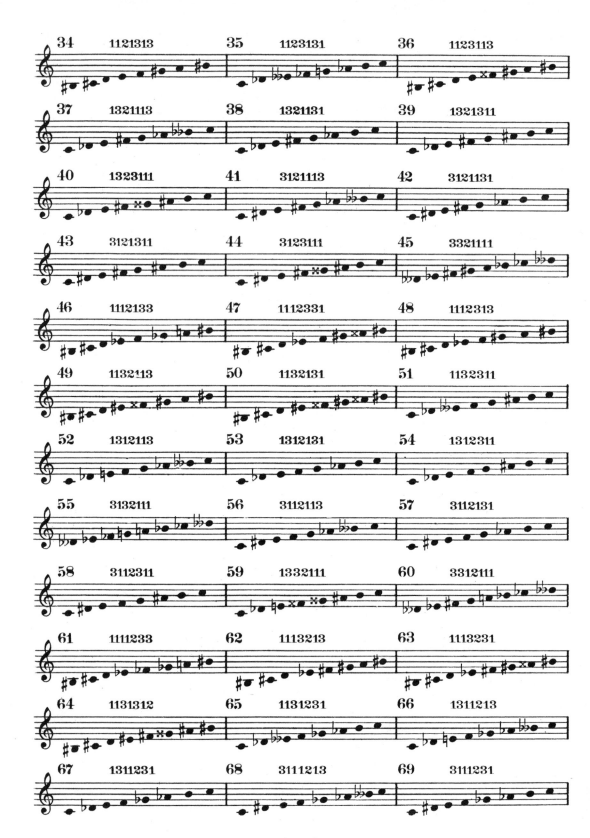

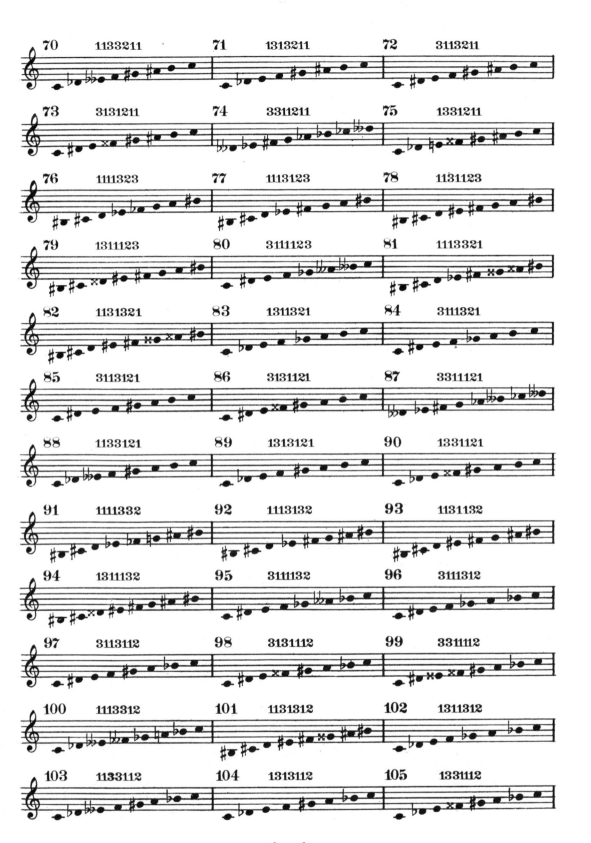

We must now investigate the chord elements on all seven tones of each of these 266 scales. It is most practical to write out all the ninth-chords, because, of the four numbers in the ninth-chord, the first two give us the triads, the first three the seventh-chords, and the ninth is the next degree in the scale.

There are 266 scales and seven ninth-chords in each scale—a total of 1,862 ninth-chords.

It would be a herculean task to write out all these chords in notes and classify them. It would be somewhat easier to calculate the chords in number-forms and write them out. Still it would be a formidable task.

Fortunately, neither notes nor form calculations are necessary. All we need is the number-form of the scale. We will take a scale generally known—the harmonic minor 2122131. Every pair of adjacent numbers is now added together, the seventh number is added to the first, and we go through the numbers once more:

2+1, 2+2, 1+3, 1+2, 1+2, 2+1, 3+1.

These numbers are now added and written out consecutively four times and divided into groups of four:

3443 3343 4433 3434 4333 4344 3334,

and in the scale they are:

| 3443 | 3343 | 4433 | 3434 | 4333 | 4344 | 3334 |
| i | ii | iii | iv | v | vi | vii |

This system applies to any seven-tone scale, and it is impossible to make a mistake provided the addition of the scale-numbers is correct. If we wish eleventh-chords, we need only add the nearest number of the adjacent group.

Wir werden jetzt die Akkorde aller sieben Stufen jeder der 266 Skalen untersuchen. Es empfiehlt sich die Akkorde als Nonenakkorde darzustellen, weil von den vier Ziffern dieses Akkordes die ersten zwei den Dreiklang, die ersten drei den Septimenakkord ergeben, und die None ist die nächste Stufe der Skala.

Es gibt 266 Skalen und sieben Nonenakkorde in jeder Skala, also eine Gesamtsumme von 1,862 Nonenakkorden.

Es würde eine Herkulesarbeit sein, alle diese Akkorde in Noten aufzuschreiben und zu klassifizieren; es würde einigermassen leichter sein, die Akkorde in Zahlenformeln zu berechnen und aufschreiben; doch würde es ein ungeheures Unternehmen sein.

Glücklicherweise sind weder Noten noch Zahlenberechnungen nötig. Alles, was wir brauchen, ist die Zahlenformel der Skala. Wir nehmen eine allgemein bekannte Tonleiter (das harmonische Moll) 2122131. Zwei benachbarte Ziffern werden addiert; die siebente Ziffer wird zur ersten addiert und wir gehen auf diese Weise die ganze Skala noch mal durch:

2+1, 2+2, 1+3, 1+2, 1+2, 2+1, 3+1.

Diese Ziffern werden jetzt viermal aufgeschrieben und eingeteilt in Gruppen von je vier:

3443 3343 4433 3434 4333 4344 3334,

und in der Skala finden wir sie auf den Stufen:

| 3443 | 3343 | 4433 | 3434 | 4333 | 4344 | 3334 |
| i | ii | iii | iv | v | vi | vii |

Dieses System kann auf irgend eine siebentonige Skala angewandt werden, und Fehler sind ausgeschlossen, falls die Addition der Skalen-Ziffern genau ist. Falls wir Undecimenakkorde darstellen wollen, brauchen wir nur die erste Ziffer von der benachbarten Gruppe hinzuzufügen.

Total result of these ninth-chords with permutation:

3335 in permutation 4
3334 in permutation 4
3344 in permutation 6
3444 in permutation 4
2444 in permutation 4
2344 in permutation 12
2335 in permutation 12
2345 in permutation 24

———

70

These chords have already been discussed.

Gesamtübersicht dieser Nonenakkorde mit Permutation:

3335 durch Permutation 4
3334 durch Permutation 4
3344 durch Permutation 6
3444 durch Permutation 4
2444 durch Permutation 4
2344 durch Permutation 12
2335 durch Permutation 12
2345 durch Permutation 24

———

70

Diese Akkorde wurden schon früher behandelt.

NEW CHORDS

From the harmonic minor:

2452 in permutation 12
3345 in permutation 12
2355 in permutation 12
2445 in permutation 12
2245 in permutation 12

—
60

From the Hungarian scale:

2236 in permutation 12
2346 in permutation 24
2256 in permutation 12
2336 in permutation 12
2255 in permutation 6
2235 in permutation 12

—
78

Grand total, 208.

The double-augmented third (6) appears here.

In a previous work by this author (*Contributions to Theory* [Berlin: Sulzbach, 1930]) there is a section dealing with altered augmented eleventh-chords derived by permutation from an altered whole-tone scale; eighteen of these chords appear in the permutation of the seven-tone Hungarian scale.

A detailed treatment of the 138 new ninth-chords belongs to the future—probably the distant future. It would be most natural to seek the resolution of them in the scales where they belong. But we are not adjusted to such scales; it has taken centuries to become adjusted to two scales, and here we are dealing with 257 new scales.

An attempt will be made to show that the chords are not entirely without resolution-possibilities even in our present diatonic system.

NEUE AKKORDE

Aus der harmonischen Moll-Tonleiter:

2452 durch Permutation 12
3345 durch Permutation 12
2355 durch Permutation 12
2445 durch Permutation 12
2245 durch Permutation 12

—
60

Aus der ungarischen Tonleiter:

2236 durch Permutation 12
2346 durch Permutation 24
2256 durch Permutation 12
2336 durch Permutation 12
2255 durch Permutation 6
2235 durch Permutation 12

—
78

Totalsumme, 208 verschiedene Nonenakkorde.

Die doppelt erhöhte Terz (6) erscheint hier.

In einem früheren Werke des Verfassers (*Beiträge zur Theorie* [Berlin: Verlag W. Sulzbach, 1930]) ist eine Abteilung, welche alterierte übermässige Undecimenakkorde bringt, abgeleitet durch Permutation von einer alterierten ganztonigen Skala; achtzehn dieser Akkorde erscheinen durch Permutation der siebentonigen ungarischen Skala.

Eine eingehende Behandlung dieser 138 neuen Nonenakkorde bleibt der Zukunft vorbehalten—wahrscheinlich einer sehr entfernten Zukunft. Das Natürlichste wäre, ihre Auflösung in den Skalen zu suchen, denen sie entnommen sind. Aber wir sind noch nicht auf solche Skalen eingestellt; zwei Skalen teilweise zu erschöpfen war eine Aufgabe von etlichen Jahrhunderten, und jetzt haben wir es mit 257 neuen Skalen zu tun.

Ein Versuch wird unternommen werden zu zeigen, dass diese Akkorde nicht gänzlich ohne Auflösungsmöglichkeiten sind, selbst in unserm gegenwärtigen diatonischen System.

A NEW SCALE DEVELOPED AS PIANO STUDY

Piano students may find the new scales useful and interesting to practice as a welcome change from the traditional major and minor scales.

As they are all seven-tone scales the fingering will work out on the same principle.

Scale 2111133 is selected and changed into a more familiar orthography.

This is only one-twelfth of one scale, because it can be transposed into eleven other keys.

EINE NEUE SKALA ALS KLAVIERÜBUNG ENT- WICKELT

Für Pianisten dürften die neuen Skalen nützlich und interessant sein als eine willkommene Variante von der traditionellen Dur und Moll Skala.

Da es sich um siebentonige Skalen handelt, wird der Fingersatz nach bestehendem System gebildet.

Die Skala 2111133 wird gewählt und die Orthographie in bequemerer Lesart dargestellt.

Dies ist nur ein Zwölftel einer einzigen Skala, weil sie nach elf anderen Tonarten transponiert werden kann.

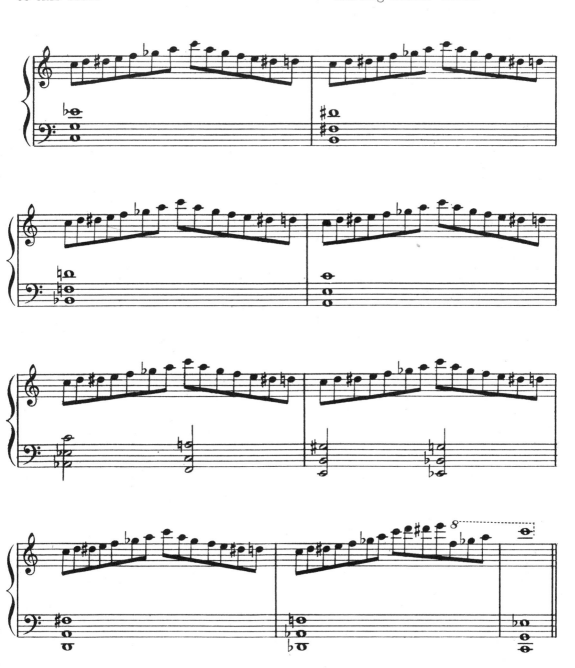

THE SPECTRUM COLORS AND THE MODULATION FORMS

The modulation forms can also be applied to the spectrum colors. Concerning the number of these, there seems to be a difference of opinion. According to the observations of Dr. A. Michelson (famous physicist, for many years connected with the University of Chicago), there are twelve colors in the spectrum. Others say there are ten.

It is outside the province of this writer to express any opinion in this matter. But if it is assumed that Dr. Michelson is right, there appears to be a parallel between the number of colors and the number of half-tones.

The writer is well aware of some of the numerous attempts to associate certain tones with certain colors, e.g., C *is* red, C *is* green, C *is* yellow, etc. Such assertions are made out of whole cloth and have no logical basis; they are the fruit of individual reaction.

As a primary condition for association of color and tone we must have a first tone and a first color. As far as music is concerned, the matter is not so complicated. Most of our literature since Bach is based upon two scales: major and minor. As the minor scale is only an altered major scale, we may omit it in our argument and deal only with the major scale. This scale in number-form is 2212221; there is only one scale which can be produced according to this number-form without resorting to chromatic alterations—C major—which therefore may be called the *first scale*. As C is the first tone of this scale we may be justified in calling C the *first tone* in music.

DIE SPEKTRALFARBEN UND DIE MODULATIONS-FORMELN

Die Modulationsformeln können auch auf die Spektralfarben angewandt werden. Über die Zahl dieser Farben herrscht Meinungsverschiedenheit. Nach den Beobachtungen von Dr. A. Michelson (berühmter Physiker, langjähriger Professor an der Universität Chicago) gibt es zwölf Spektralfarben, während andere Autoritäten nur zehn angeben.

Es ist nicht die Sache des Verfassers eine eigene Meinung zu äussern. Angenommen, dass Dr. Michelson's Behauptung zu Recht besteht, wäre das Verhältnis der Zahl der Farben und der Zahl der zwölf Halbtöne hergestellt.

Der Verfasser ist sich einiger der mannigfachen Versuche, gewisse Töne mit gewissen Farben zu verbinden, bewusst, z.B. C *ist* rot, C *ist* grün, C *ist* gelb, etc. Solche Behauptungen sind natürlich aus der Luft gegriffen; sie haben keine wissenschaftliche Begründung und sind Resultate persönlicher Reaktion.

Als erste Bedingung für Verbindung von Farbe und Ton müssen wir einen ersten Ton und eine erste Farbe haben. Was die Musik betrifft ist die Sache einfach. Fast die gesamte Literatur seit Bach basiert auf zwei Skalen (Dur und Moll). Die Moll-Tonleiter ist weiter nichts als eine alterierte Dur Skala, weshalb wir sie in unseren Betrachtungen auslassen können und uns nur mit der Dur-Tonleiter beschäftigen. Die Zahlenformel dieser Skala ist 2212221; nur eine einzige Skala kann nach dieser Formel gebildet werden ohne zu chromatischen Aenderungen Zuflucht nehmen zu müssen—C Dur. Diese mag daher als erste Skala bezeichnet werden. C ist der erste Ton dieser Skala, weshalb wir vielleicht berechtigt sind, C als den ersten Ton der Musik zu betrachten.

Whether a similar argument can be applied to colors must be left to others to decide. Whether there is such a thing as a first color is a problem for people who specialize in theory of colors. All the information this writer has obtained so far is that the first color is yellow; others say it is red and point to the rainbow, where the first color is red.

If it can be stated positively which color is No. 1 (where the spectrum begins), then there is a possibility of associating the first color with the first tone, and then the problem is probably solved, because all the other colors are a matter of proportion.

We know that the nearest related foreign tone to a given tone is the fifth, because it is the second overtone; the octave (the first overtone) is the same tone, and it is the first multiple of vibrations of the basic tone. The fifth is the second multiple and therefore the second overtone; e.g., if we say that C has eight vibrations, c one octave higher will have 8×2 vibrations, g 8×3, $\bar{c}$ 8×4, $\bar{e}$ 8×5, $\bar{g}$ 8×6, etc. The nearest relation of foreign tones is the fifth, and this takes us through the circle of fifths.

We know from elementary harmony that the major triads on C and on D are not directly related because D major has no C and C major has no F♯. But they are indirectly related through G major, the scale between them, because G major scale has a major triad on C and a major triad on D.

If we select at random a color in the spectrum—e.g., yellow and its two adjacents: yellow-green and yellow-orange—we will find that in mixing the two adjacents, yellow-green and yellow-orange, we will produce yellow. Any two colors separated by another color will produce the color which separates them. The exact colortone is only a matter of proportion. The conclusion is that the relations of colors in the spectrum corre-

Ob ähnliche Argumente auf Farben angewandt werden können, mag dahingestellt bleiben. Ob es eine erste Farbe gibt, mögen Wissenschaftler der Farbe entscheiden. Alles was der Verfasser in Erfahrung bringen konnte, ist, dass die erste Farbe gelb ist, während andere rot angeben und als Beweis den Regenbogen anführen, dessen erste Farbe rot ist.

Wenn es positiv nachgewiesen werden kann, welche Farbe No. 1 ist, wo das Spektrum anfängt, dann haben wir eine Möglichkeit für eine Verbindung der ersten Farbe mit dem ersten Ton, und dann ist das Problem wahrscheinlich gelöst, weil die andern Farben sich dann logisch ergeben.

Wir wissen, dass der nächste fremde Ton zu einem gegebenen die Quinte ist, weil diese der zweite Oberton ist; die Oktave (der erste Oberton) ist derselbe Ton und ist das erste Multiplikat der Schwingungen des Grundtones. Die Quinte ist das zweite Multiplikat und deshalb der zweite Oberton; z.B. falls wir sagen, dass C acht Schwingungen hat, wird das kleine c (eine Oktave höher) 8×2 Schwingungen haben, g 8×3, $\bar{c}$ 8×4, $\bar{e}$ 8×5, $\bar{g}$ 8×6, etc. Die nächste Verwandtschaft fremder Töne ist die Quinte, und daher kommt der Quintenzirkel.

Wir wissen aus der Elementarharmonielehre, dass die Durdreiklänge C und D keine direkte Verwandtschaft besitzen, weil D Dur kein C hat und C Dur kein Fis. Aber sie sind indirekt verwandt durch G Dur, die Skala zwischen den beiden, weil G Dur einen Durdreiklang auf C und D besitzt.

Wenn wir aufs Geratewohl eine Spektralfarbe aussuchen, z.B. gelb und ihre zwei Nachbarfarben (gelb-grün und gelb-orange), so finden wir, dass durch Vermischung von gelb-grün und gelb-orange gelb entsteht. Durch Vermischung zweier Farben, durch eine dritte getrennt, entsteht die Farbe, welche sie trennt. Die genaue Farbenschattierung ist relativ. Nach dieser Folgerung können wir feststellen, dass die Verwandtschaft der Spektralfarben dem Quintenzirkel in der

spond to the circle of fifths in music. If we say that C is the first tone and yellow is the first color in the twelve-color spectrum and we go from yellow through red to violet, blue, and green, the arrangement will be:

0   C—yellow
1   G—yellow-orange
2   D—orange
3   A—red-orange
4   E—red
5   B—red-violet
6   F#—violet
7   Db—blue-violet
8   Ab—blue
9   Eb—blue-green
10 Bb—green
11 F—yellow-green

If red is the first color (0), start here and proceed either up or down, either right or left, in the circle.

In order to make the colors serviceable for the modulation forms, it will be necessary to change the circle of fifths into a chromatic circle of half-tones. This change is produced by exchanging 1 and 7, 3 and 9, 5 and 11. The chromatic circle is now:

0   C—yellow
1   Db—blue-violet
2   D—orange
3   Eb—blue-green
4   E—red
5   F—yellow-green
6   F#—violet
7   G—yellow-orange
8   Ab—blue
9   A—red-orange
10 Bb—green
11 B—red-violet

These colors can now be arranged according to any modulation form. As any tone may be the starting-point (0), so also the colors.

In music we do not observe any marked difference between a statement in C major and the same statement transposed to D flat major; but in colors the differ-

Musik entspricht. Wenn wir sagen, dass C der erste Ton ist und gelb die erste Farbe in dem Zwölf-Farbenspektrum, und wir von gelb durch rot nach violett, blau und grün uns bewegen, so wird die Anordnung folgende sein:

0   C—gelb
1   G—gelb-orange
2   D—orange
3   A—rot-orange
4   E—rot
5   H—rot-violett
6   Fis—violett
7   Des—blau-violett
8   As—blau
9   Es—blau-grün
10 B—grün
11 F—gelb-grün

Falls rot die erste Farbe ist (0), fangen wir da an und bewegen uns entweder auf oder ab, rechts oder links in diesem Zirkel.

Um diese Farben nutzbar zu machen für die Modulationsformeln, wird es nötig sein den Quintenzirkel in einen chromatischen zu verwandeln, in einen Zirkel mit Halbtonfortschreitung. Dieser Wechsel findet statt, indem 1 und 7, 3 und 9, 5 und 11 gewechselt werden. Der chromatische Zirkel ist jetzt:

0   C—gelb
1   Des—blau-violett
2   D—orange
3   Es—blau-grün
4   E—rot
5   F—gelb-grün
6   Fis—violett
7   G—gelb-orange
8   As—blau
9   A—rot-orange
10 B—grün
11 H—rot-violett

Diese Farben können jetzt nach irgend einer Modulationsformel arrangiert werden. Ebenso wie irgend ein Ton den Anfang (0) bilden kann, so die Farben.

In der Musik machen wir nicht viel Unterschied zwischen einem Satz in C Dur und demselben nach Des Dur transponiert; aber in Farben ist der Unterschied ungeheuer gross. Als ein Beispiel

ence is enormous. As an example we will select the colors corresponding to the keys of the *D Flat Nocturne*, Op. 27, No. 2, by Chopin and in a parallel column the colors in its transposition into D major as Wilhelmi has done in his transcription for violin and piano:

D flat major	D major
blue-violet	orange
green	red-violet
blue-green	red
blue-violet	orange
red-orange	green
blue-violet	orange
violet	yellow-orange
green	red-violet
blue-violet	orange

and the transcription is almost the diametrical opposite of the original key. All the "cold" colors in D flat are "warm" in D; the one "warm" color in D flat is "cold" in D.

It is not to be wondered at that the eye should be much keener than the ear; the eye has been adjusted to colors in nature as far back as humanity can be traced, while we have been associated with organized sounds only a few centuries.

Our modulation work-form with twelve different starting-points will, in colors, produce twelve different results. As we have 7,708 work-forms, the number of arrangements of the colors is 7,708 × 12 = 92,496.

wollen wir die Farben wählen, welche den Modulationen in der *Des dur Nocturne* von Chopin entsprechen und in einer Parallel-Reihe die Farben in der Transposition nach D Dur, wie es Wilhelmi in seiner Transcription für Violine und Klavier gemacht hat:

Des dur	D dur
blau-violett	orange
grün	rot-violett
blau-grün.	rot
blau-violett	orange
rot-orange	grün
blau-violett	orange
violett	gelb-orange
grün	rot-violett
blau-orange	orange

und die Transposition ist ungefähr der schroffste Gegensatz zur Originaltonart —Des Dur. Alle "kalten" Farben in Des sind "warm" in D.

Wir brauchen uns nicht zu wundern, dass das Auge ein schärferes Organ ist als das Ohr; das Auge ist auf Farben in der Natur eingestellt seit Entstehung der Menschheit, während das Ohr sich mit geordnetem Tonsystem erst seit einigen Jahrhunderten beschäftigt hat.

Unsere Modulations-Arbeitsformel mit zwölf verschiedenen Anfangspunkten wird in Farben zwölf verschiedene Resultate ergeben. Da wir 7,708 Arbeitsformeln haben, so ergeben sich als Farben-Arrangements 7,708 × 12 = 92,496.

APPENDIX

By ERNEST BLOOMFIELD ZEISLER

CONGRUENCE

When the difference between two numbers c and d is divisible by b, we say that c is congruent to d modulo b, and write this

$$c \equiv d \ (b), \text{ or } c \underset{b}{\equiv} d \ . \tag{1}$$

Let

$$w_1, w_2, w_3, \ldots, w_{b-1}, w_b \tag{2}$$

be a permutation of the first $b \geqq 2$ integers

$$1, 2, 3, \ldots, b-1, b \ . \tag{3}$$

The *interval* from w_i to w_j is defined as the smallest positive integer which is congruent modulo b to $w_j - w_i$, that is, as $w_j - w_i$ if $w_j \geqq w_i$ and as $w_j - w_i + b$ if $w_j < w_i$. With this convention we may say that the interval from w_i to w_{i+1} is k_i where $k_i \equiv w_{i+1} - w_i \ (b)$, and $1 \leq k_i \leq b-1$.

WORK-FORMS

By a work-form we mean a permutation (2) of the integers (3) such that the intervals

$$k_1, k_2, k_3, \ldots, k_{b-2}, k_{b-1} \tag{4}$$

are all distinct. Hence each work-form (2) determines a sequence (4) which is called a *key-form*, which is a permutation of the integers

$$1, 2, 3, \ldots, b-2, b-1 \ . \tag{5}$$

It follows from the definition of k_i that

$$\begin{cases} w_2 \equiv w_1 + k_1 \ (b) \\ w_3 \equiv w_1 + k_1 + k_2 \ (b) \\ \cdot \quad \cdot \quad \cdot \quad \cdot \quad \cdot \quad \cdot \\ w_{i+1} \equiv w_1 + k_1 + k_2 + \cdots + k_i \ (b) \\ \cdot \quad \cdot \quad \cdot \quad \cdot \quad \cdot \quad \cdot \\ w_b \equiv w_1 + k_1 + k_2 + \cdots + k_{b-1} \ (b) \end{cases} \tag{6}$$

Let K_i be defined as

$$K_i = k_1 + k_2 + \cdots + k_i \ ; \tag{7}$$

Then by (6)

$$w_{i+1} \equiv w_1 + K_i \ (b) \ . \tag{8}$$

[156]

Hence each key-form (4) determines a work-form (2) if we choose a w_1.
By (8) we have

$$w_{i+1} \equiv w_1 + K_i \; (b)$$
$$w_{j+1} \equiv w_1 + K_j \; (b) \; ; \qquad\qquad (9)$$

hence

$$w_{j+1} - w_{i+1} \equiv K_j - K_i \; (b) \; . \qquad\qquad (10)$$

If $i \neq j$ then w_{i+1} and w_{j+1} are distinct, so that

$$K_j - K_i \not\equiv 0 \; (b) \; \text{for} \; i \neq j \; . \qquad\qquad (11)$$

Hence *a key-form is a permutation* (4) *of the integers* (5) *with the property* (11).

THE PROBLEM

The problem is to find all the work-forms. From what has been shown above, this may be accomplished by finding all the key-forms.

Theorem I. There exists no key-form when b is odd.

$$K_{b-1} = k_1 + k_2 + \cdots + k_{b-1} = 1 + 2 + 3 + \cdots + (b-1) = \frac{b(b-1)}{2} \; ; \qquad\qquad (12)$$

when b is odd, then $(b-1)$ is even, so that K_{b-1} is divisible by b, which contradicts (11) with $j = b-1$ and $i = 0$; hence there is no key-form.

Theorem II. For every even b there exists the key-form

$$k_{2p+1} = 2p + 1, \; k_{2p} = b - 2p \; , \qquad\qquad (13)$$

and *the middle element is* $k_{\frac{b}{2}} = \dfrac{b}{2}$.

To make the form (13) more concrete we may write it:

$$1, b-2, 3, b-4, 5, b-6, \ldots, b-5, 4, b-3, 2, b-1 \; ,$$

that is, we arrange the odd numbers in natural order from left to right and place between them the even numbers in natural order but from right to left. This is a key-form if, and only if, it satisfies (11). To investigate this we introduce two lemmas:

Lemma 1.—The sum of the first n even numbers is

$$S_n^e = n(n+1) \; . \qquad\qquad (14)$$

This is clear since the sum of the first n even numbers is twice the sum of the first n numbers, $1 + 2 + 3 + 4 + \cdots + n$, and the latter sum is $\dfrac{n(n+1)}{2}$.

Lemma 2.—The sum of the first n odd numbers is

$$S_n^o = n^2 \; . \qquad\qquad (15)$$

This is clear since each number of (15) is one less than the corresponding number of (14), so that

$$S_n^o = S_n^e - n = n^2 + n - n = n^2 .$$

Let us now evaluate K_i. If i is odd, than K_i is the sum of the *first* $\dfrac{i+1}{2}$ odd numbers of (5) and the *last* $\dfrac{i-1}{2}$ even numbers of (5). By (15) the sum of the first $\dfrac{i+1}{2}$ odd numbers is $\left(\dfrac{i+1}{2}\right)^2$. The sum of the last $\dfrac{i-1}{2}$ even numbers of (5) is clearly the sum of all the $\dfrac{b-2}{2}$ even numbers of (5) minus the sum of the first $\dfrac{b-i-1}{2}$ even numbers; and by (14) this is

$$S_{\frac{b-2}{2}}^e - S_{\frac{b-i-1}{2}}^e = \frac{b-2}{2} \cdot \frac{b}{2} - \frac{b-i-1}{2} \cdot \frac{b-i+1}{2} = \tfrac{1}{4}(2ib - 2b - i^2 + 1) = \frac{b(i-1)}{2} - \frac{i^2-1}{4} ;$$

since i is odd, $(i-1)$ is even, so that $\dfrac{b(i-1)}{2}$ is an integral multiple of b; hence this last quantity is congruent modulo b to $-\dfrac{i^2-1}{4}$; hence when i is odd,

$$K_i \underset{b}{\equiv} \left(\frac{i+1}{2}\right)^2 - \frac{i^2-1}{4} = \frac{i+1}{2} .$$

If i is even, then

$$K_i = K_{i-1} + k_i \underset{b}{\equiv} \frac{i}{2} + b - i \underset{b}{\equiv} -\frac{i}{2} ;$$

for since i is even, $(i-1)$ is odd, and by the foregoing we have $K_{i-1} \underset{b}{\equiv} +\dfrac{i}{2}$, and $k_i = b - i$ by (13) when i is even. Hence

$$\text{when } i \text{ is odd, } K_i \underset{b}{\equiv} \frac{i+1}{2} \text{ ; and when } i \text{ is even, } K_i \underset{b}{\equiv} -\frac{i}{2} . \tag{16}$$

Case 1: i and j both even.—By (16) $K_j - K_i \equiv -\dfrac{j-i}{2}$, which is not zero when $i \neq j$.

Case 2: i even and j odd.—By (16) $K_j - K_i \equiv \dfrac{j+1}{2} + \dfrac{i}{2}$; now i and j are both less than b and they are distinct; hence $i+j$ is not more than $2b-3$, so that $i+j+1$ is not more than $2b-2$; hence $\dfrac{j+i+1}{2}$ is less than b and cannot be congruent to b, since it is obviously not zero.

Case 3: i odd and j even.—By (16) $K_j - K_i \equiv -\dfrac{j}{2} - \dfrac{i+1}{2}$; this is the negative of the result in Case 2 and is therefore also not divisible by b.

Case 4: i and j both odd.—By (16) $K_j - K_i \equiv \dfrac{j+1}{2} - \dfrac{i+1}{2} = \dfrac{j-i}{2}$, which is not zero when $i \neq j$.

Hence in every case (11) is satisfied, so that (13) is a key-form, as was to be proved. As to the second part of the theorem: by (13) k_b is either $\dfrac{b}{2}$ or $b - \dfrac{b}{2}$, so that in every case $k_{\frac{b}{2}} = \dfrac{b}{2}$. Hence Theorem II is proved.

Theorem III. For every work-form it is true that $w_b - w_1 \equiv \dfrac{b}{2}$ (b).

For by (8), $w_b - w_1 \equiv K_{b-1}$; since b is even $(b-1)$ is odd, so that by (12) $K_{b-1} \equiv \dfrac{b(b-1)}{2} \equiv \dfrac{b}{2}$. Thus the last entry of the work-form is always removed from the first entry by the interval $\dfrac{b}{2}$.

Theorem IV. The first and the last intervals in a work-form cannot be $\dfrac{b}{2}$ *(unless $b=2$).*

For if the first interval were $\dfrac{b}{2}$, then it follows from Theorem III that w_2 and w_b both are removed from w_1 by the same interval and hence $w_2 = w_b$, which is not possible if $b \neq 2$; the argument is similar for the last interval.

The general key-form.—We have seen (Theorem II) that for every even b there exists a key-form, and that its first and last elements are not $\dfrac{b}{2}$ (Theorem IV); hence every key-form may be written

$$K: A, \frac{b}{2}, C \tag{17}$$

where A and C are sequences with at least one member in each. Let us define A^t as the sequence A in reverse order, C' as C in reverse order, and K' as K in reverse order, so that K' is $C', \dfrac{b}{2}, A'$.

Theorem V. For every key-form the reverse is also a key-form.

Let the key-form K be the sequence (4). Then K' is the sequence $k'_1, k'_2, k'_3, \ldots, k'_{b-1}$ where $k'_i = k_{b-i}$. We define K'_i in terms of the k' in the same way (7) as K_i in terms of the k. Then clearly K'_i is the sum of the first i elements of K' and therefore the sum of the last i elements of K, that is, $K'_i = K_{b-1} - K_{b-i-1}$. Therefore

$$K'_j - K'_i = (K_{b-1} - K_{b-j-1}) - (K_{b-1} - K_{b-i-1}) = K_{b-i-1} - K_{b-j-1} \; ;$$

for $i \neq j$ the right member of this equation is distinct from zero, so that K' satisfies (11) as well as does K; hence K' is a key-form.

Theorem VI. If $A, \dfrac{b}{2}, C$ is a key-form then so is $A', \dfrac{b}{2}, C'$.

Suppose $A', \dfrac{b}{2}, C'$ is not a key-form. Then it does not satisfy (11); therefore there is a sequence of consecutive elements in $A', \dfrac{b}{2}, C'$ whose sum is divisible by b, for the left member of the inequality (11) is simply the sequence of consecutive elements of (4) from the $(i+1)$st through the jth.

Let S be the consecutive sequence of $A', \dfrac{b}{2}, C'$ which is a multiple of b.

Case 1: S is a part of A'.—Then its reverse, S', is a consecutive sequence of A, and its sum is a multiple of b, since reversing the order of a set of numbers does not alter their sum. Hence there is a consecutive sequence, S', of $A, \dfrac{b}{2}, C$ which is a multiple of b, so that $A, \dfrac{b}{2}, C$ is not a key-form.

Case 2: S is a part of C.—The same argument as in Case 1 applies here also.

Case 3: S includes the element $\dfrac{b}{2}$.—Let S be $A'_2, \dfrac{b}{2}, C'_2$, where A'_2 is the last part of A' and C'_2 is the first part of C'. Then we may write $A', \dfrac{b}{2}, C'$ as $A'_1, A'_2, \dfrac{b}{2}, C'_2, C'_1$ and $A, \dfrac{b}{2}, C$ as $A_2, A_1, \dfrac{b}{2}, C_1, C_2$. Now we are given that $A'_2 + \tfrac{1}{2}b + C'_2$ is a multiple of b; hence $A_2 + \tfrac{1}{2}b + C_2$ is the same multiple of b; but by (12) we

know that $A_2+A_1+\frac{1}{2}b+C_1+C_2$ is $\frac{1}{2}b(b-1)$ since it is the sum of the first $b-1$ integers, regardless of whether it is or is not a key-form. When b is even, $(b-1)$ is odd, so that this sum is an odd multiple of $\frac{1}{2}b$. But $A_2+\frac{1}{2}b+C_2$ is a multiple of b and therefore an even multiple of $\frac{1}{2}b$; hence $A_2+A_1+\frac{1}{2}b+C_1+C_2$ $-(A_2+\frac{1}{2}b+C_2)$ equals an odd multiple of $\frac{1}{2}b$ minus an even multiple of $\frac{1}{2}b$, which gives A_1+C_1 is an odd multiple of $\frac{1}{2}b$. Hence $A_1+\frac{1}{2}b+C_1$ is an even multiple of $\frac{1}{2}b$ and therefore is a multiple of b; but this is a consecutive sequence of $A, \dfrac{b}{2}, C$, so that the form $A, \dfrac{b}{2}, C$ is not a key-form.

Thus in every case we see that, if $A', \dfrac{b}{2}, C'$ is not a key-form, then neither is $A, \dfrac{b}{2}, C$; consequently, if $A, \dfrac{b}{2}, C$ is a key-form, then so is $A', \dfrac{b}{2}, C'$, which is the theorem.

Corollary. The total number of key-forms is a multiple of four if b is greater than 4.—If b is odd, then there exists no key-form (Theorem I), so that the total number of key-forms is zero, which is a multiple of 4.

Let b be even. Since b is greater than 4, we know that $b-1$ is greater than 3; by (17) every key-form can be written $A, \dfrac{b}{2}, C$, where A and C each contains at least one member. Since the total number of elements in this form is greater than 3, at least one of the sequences A or C contains more than one element. Suppose A contains more than one element. Then A' is not the same as A; but by Theorem VI $A', \dfrac{b}{2}, C'$ is a key-form, and this is distinct from $A, \dfrac{b}{2}, C$ since A' and A are distinct; hence for every key-form $A, \dfrac{b}{2}, C$ there are three additional distinct key-forms: $C, \dfrac{b}{2}, A$ and $A', \dfrac{b}{2}, C'$ and $C', \dfrac{b}{2}, A'$; hence the total number of key-forms is a multiple of 4.

Theorem VII. The natural sequence (5) is a key-form if, and only if, b is a power of 2.

As we saw before, (5) is a key-form if, and only if, it satisfies (11). Now in (5) the elements are $k_i=i$; consequently $K_i=\frac{1}{2}i(i+1)$, so that

$$K_j-K_i=\tfrac{1}{2}(j^2+j-i^2-i)=\tfrac{1}{2}(j-i)(j+i+1) . \tag{18}$$

Suppose b is a power of 2, that is, $b=2^m$.

Case 1: i and j both even or both odd.—Hence $i+j+1$ is odd and does not contain the factor 2. Also $j-i$ is surely less than b, so that $\frac{1}{2}(j-i)$ contains the factor 2 at most $m-2$ times, so that the quantity (18) contains the factor 2 at most $m-2$ times and therefore cannot be a multiple of b unless it is zero; but $j-i$ is not zero if $j\neq i$, and $j+i+1$ is surely not zero, so that the quantity (18) is not a multiple of b.

Case 2: i odd and j even, or i even and j odd.—Now $j+i+1$ is even; but since i and j are both less than b, it follows that $j+i+1$ is less than $2b$, so that $\frac{1}{2}(j+i+1)$ is less than b and contains the factor 2 less than m times; $j-i$ is an odd number, so that it does not contain the factor 2 at all; hence the quantity (18) contains the factor 2 less than m times and so is not divisible by b unless it is zero; that it is not zero follows as in Case 1.

Thus, if b is a power of 2, then (18) is not divisible by b, and (5) is a key-form. Conversely, if (5) is a key-form, then b is a power of 2. For suppose b is not a power of 2; then we may factor out all the powers of 2 that we can and write $b=2^m(2n+1)$, with $n\neq0$.

Case 1: n less than 2^m.—Then the sequence

$$2^m-n, 2^m-n+1, \ldots, 2^m-1, 2^m, 2^m+1, \ldots, 2^m+n-1, 2^m+n \tag{19}$$

is a consecutive sequence in (5), because its first element 2^m-n is at least one since n is less than 2^m and its last element 2^m+n is less than 2^m+2^m for the same reason, and hence is less than b. The sum of the sequence (19) is obviously

$$(2^m-n+2^m+n)+(2^m-n+1+2^m+n-1)+ \cdots +(2^m-1+2^m+1)+2^m ,$$

for we have simply rearranged the terms, grouping the last with the first, the next to the last with the second, and so on—a procedure which does not alter the sum. The last sum is $2n(2^m)+2^m$, which is $2^m(2n+1)$ or b; consequently, the sum of the sequence (19) is divisible by b, so that (5) is not a key-form.

Case 2: n greater than 2^m.—Then the sequence

$$n-2^m+1, \ldots, n, n+1, \ldots, n+2^m , \tag{20}$$

is a consecutive sequence in (5), because its first element is at least one, since n is greater than 2^m and its last element $n+2^m+1$ is at most $2n$ and hence is less than b. The sum of the sequence (20) is

$$(n-2^m+1+n+2^m)+(n-2^m+2+n+2^m-1)+ \cdots +(n-1+n+2)+(n+n+1) ,$$

where again we have grouped the last with the first, etc. This last sum is readily seen to be $(2n+1)2^m$ or b, so that again (5) is not a key-form.

Case 3: n equals 2^m.—The sequence $1, 2, 3, \ldots , 2n$ is a consecutive sequence in (5), since $2n$ is less than b; the sum of this sequence is $\frac{1}{2}2n(2n+1)=n(2n+1)=2^m(2n+1)=b$; hence again (5) is not a key-form.

Thus if b is not a power of 2 the sequence (5) is not a key-form; hence if (5) is a key-form b is a power of 2.

Classification of key-forms.—If in a key-form the middle element is $\frac{1}{2}b$, then the form is called a *central* key-form; otherwise it is called an *acentral* key-form. By Theorem II, when b is even, there always exists a central key-form. If b is greater than 4, then by the Corollary we need to find only one-fourth of all the key-forms in order to have them all, since each one *generates* three more. To find a complete set of generating key-forms we may proceed as follows:

1. Find all the central key-forms with the element 1 in the first half of A or in its middle.

2. Find all the acentral key-forms with B longer than A and with the element 1 in the first half of A or in its middle.

3. Find all the acentral key-forms with B longer than A and with the element 1 in the first half of B or in its middle.

Complete sets of generating key-forms for $b=2, 4, 6, 8, 10$ are as follows:

$b=2:$

　　　　1

$b=4:$

　　　123

$b=6:$

　　14325

$b=8:$

CENTRAL	ACENTRAL
1234567	1643752
1634527	3241576
5174632	7245136
	6542137

$b=10:$

CENTRAL

$k_1=1:$	126357489	$k_2=1:$	312654897
	124753689		318456297
	124759863		813456792
	174258639		
	138654279		
	183654729		
	176852439		
	176859342		

$k_4 = 5$:	129574836	$k_3 = 5$:	125978436	$k_2 = 5$:	153869247
	162548793		135869427		158324967
	184597263		175948326		157632498
			185369427		
	216534789				351267849
			765123849		851327694
	498512367		985132674		851396724
	367512498		485132679		751623489
	867512493		265134879		251678439
	926513847		935162784		851762349
	427513968		435162789		351872694
	472518963				351896274
			475218693		
	347521689		865714293		657132948
	849571326		365714298		256143879
	342571689		425718693		756148329
					657149238
	386524179		975241683		952168743
	638592147		925741683		453186927
					653814297
					259714386
					759214836
					953416872
					452618793
					452781693
					957241863

Base b	GENERATING KEY-FORMS				TOTAL KEY-FORMS
	Central		Acentral	Total	
	$k_1 = 1$	Total			
2............	I	I	O	I	I
4............	I	I	O	I	2
6............	I	I	O	I	4
8............	2	3	4	7	28
10............	8	I I	55	66	264

The problem of the number of key-forms for the general base b is a problem in partitions and probably admits of no formula.

DATE DUE